# Science and Fiction

## Science and Fiction – A Springer Series

This collection of entertaining and thought-provoking books will appeal equally to science buffs, scientists and science-fiction fans. It was born out of the recognition that scientific discovery and the creation of plausible fictional scenarios are often two sides of the same coin. Each relies on an understanding of the way the world works, coupled with the imaginative ability to invent new or alternative explanations—and even other worlds. Authored by practicing scientists as well as writers of hard science fiction, these books explore and exploit the borderlands between accepted science and its fictional counterpart. Uncovering mutual influences, promoting fruitful interaction, narrating and analyzing fictional scenarios, together they serve as a reaction vessel for inspired new ideas in science, technology, and beyond.

Whether fiction, fact, or forever undecidable: the Springer Series "Science and Fiction" intends to go where no one has gone before!

Its largely non-technical books take several different approaches. Journey with their authors as they

- Indulge in science speculation – describing intriguing, plausible yet unproven ideas;
- Exploit science fiction for educational purposes and as a means of promoting critical thinking;
- Explore the interplay of science and science fiction – throughout the history of the genre and looking ahead;
- Delve into related topics including, but not limited to: science as a creative process, the limits of science, interplay of literature and knowledge;
- Tell fictional short stories built around well-defined scientific ideas, with a supplement summarizing the science underlying the plot.

Readers can look forward to a broad range of topics, as intriguing as they are important. Here just a few by way of illustration:

- Time travel, superluminal travel, wormholes, teleportation
- Extraterrestrial intelligence and alien civilizations
- Artificial intelligence, planetary brains, the universe as a computer, simulated worlds
- Non-anthropocentric viewpoints
- Synthetic biology, genetic engineering, developing nanotechnologies
- Eco/infrastructure/meteorite-impact disaster scenarios
- Future scenarios, transhumanism, posthumanism, intelligence explosion
- Virtual worlds, cyberspace dramas
- Consciousness and mind manipulation

More information about this series at http://www.springer.com/series/11657

Pernille Rørth

# The Unedited

A Novel About Genome and Identity

 Springer

Pernille Rørth
Bisley, Gloucestershire, UK

ISSN 2197-1188          ISSN 2197-1196   (electronic)
Science and Fiction
ISBN 978-3-030-34623-2          ISBN 978-3-030-34624-9    (eBook)
https://doi.org/10.1007/978-3-030-34624-9

Cover art by Stephen Cohen

This Springer imprint is published by the registered company Springer Nature Switzerland AG.
The registered company address is: Gewerbestrasse 11, 6330 Cham, Switzerland

# Contents

# Part I

# 1

# The Wall

"It's just a huge fucking wall."

"Of course it is. What did you expect?" Leo says, tapping the controls. The pod swerves smoothly to the left. They start moving alongside the looming mass of concrete.

"Something a bit more sophisticated," Raphael responds. "Our latest technology to keep out the barbarians. *And* their deadly diseases."

"They probably consider us the barbarians," Eiko says under her breath. Only Ben, who is sitting next to her in the back seat, hears it. He shrugs but adds no comment of his own. He continues to look out the window. Eiko follows his gaze. The solid gray structure streaming past them is strangely mesmerizing. It goes on and on, massive, smooth and silent. It bulges out here and there, possibly accommodating something on the other side. It towers above them, ominously, when they come in close. But it is just a wall.

"Why don't we just hop over it?" Raphael asks, a few minutes later. He is sitting on the other side of Eiko. "It's not that tall."

"A shitload of laser-zappers on top," Leo answers, waving a hand in that direction. "But I'll let you out so you can find out for yourself." He smirks. "Zap, zap, pong, poor Rafi's gone."

"Fuck you," Raphael responds.

"Rafi," Celia says, "don't be so sensitive." She turns to Leo, her voice flat. "And you—don't be such a prick." He looks surprised, almost shocked. She notices and hides a miniscule smile. Switching to a lighter tone she adds "but wouldn't they be directional?"

It takes Leo a moment to realize what she is talking about. "Our zappers would be, for sure. But they've put up their own, as well." He taps the

© Springer Nature Switzerland AG 2020
P. Rørth, *The Unedited*, Science and Fiction, https://doi.org/10.1007/978-3-030-34624-9_1

dashboard display and a camera from the micro-scout zooms in. "They look almost exactly like Huang flats." He scoffs. "I wouldn't be surprised if the software is copied as well. Pathetic, really." He looks at Celia and raises an eyebrow. "I've tricked flat zappers before. Do you want me to try?"

"They'll revoke the visas if we do anything stupid," Eiko interjects before Celia can answer. "Let's just find the transit point."

"Yes, Ma'am," Leo says. Eiko does not respond.

"But you could take it up a bit—so we can see what's in there," Celia says, touching Leo lightly on the arm.

"Celia, please," Eiko says with a sigh. "Be patient."

"Come on, Eiko, aren't you curious?" Celia's tone is light and playful. She turns around and smiles at Eiko. "Let's just have a peek. The satellite photos are useless."

"Scramble shield," Leo says.

"Naturally," Celia says and turns toward him. "But this close, and from this angle, we should be able to see something, shouldn't we?"

"Probably," Leo concedes. But he continues flying low.

"I'm more curious about the people," Eiko says, with a tentative smile. "All we know is their history, their old-fashioned rules-"

"But we've never met any of them," Celia interrupts, turning back to Eiko and nodding enthusiastically. "This is so exciting!" She exclaims, then pauses. "They might be quite primitive, though. Maybe they think we-" Eiko sends her a quick look of reprimand with a side-glance to Ben. Celia mouths "sorry" and turns to face forward again.

No one says anything for while.

Raphael glances at his wrist-link for the twenty-seventh time, but resists touching it. He does not need another condescending explanation of why there is no connection inside Leo's pod. He looks out the window, instead. On his side, trees pass in a blur.

"Look!" Celia suddenly says and points to a section of the wall immediately ahead of them. "The wall has collapsed." They all turn their heads as they glide past a series of dramatic vertical cracks with piles of gray rubble below. "Well, not quite," she continues, "but I guess it *is* pretty old. It was built during the crisis, wasn't it?"

"Threatened by superstition and ignorance…" Eiko starts.

"…we found the courage to act," Raphael continues, the pitch of his voice rising to match that of their former teacher. "Knowledge and fortitude, children. Remember that! Knowledge *and* fortitude." He is a good mimic. He and Eiko giggle. Ben glances over at them and smiles.

"We certainly will remember Ms. Clifton." Eiko says, once the giggles have subsided.

"You three," Celia says, shaking her head. "Sometimes it's like you never left primary school."

"You missed the best part."

"Believe me, I had my share of-"

"Could that be it?" Leo interrupts them. "That thing up ahead, close to the wall?"

They see a rectangular structure in the distance, alternatingly darker than the wall and sparkling bright. As it grows bigger, they continue to stare at it. Their expressions range from eager to apprehensive.

"It looks about right," Celia says briskly, breaking the long silence. The building is two stories tall and directly connected to the wall. It is part concrete, part steel and glass, and looks quite new. Around it, the clearing has been widened considerably and partially landscaped. There are no parked pods or any other signs of activity.

"Taking it down, folks." Leo decelerates the pod and lands it in the gravel-covered area marked "Visitors' parking".

As soon as the pod is stationary, Raphael opens his door and jumps out. Ben and Eiko exit more slowly from the other side, grab their backpacks and walk away from the pod, their eyes fixed on the silent building. Celia also takes her time getting out. She looks around and inhales deeply. She smiles; the air is fresh and pleasant. Then she moves toward where Ben and Eiko are standing. Raphael remains near the pod, fidgeting with his wrist-link. Leo is the last person to exit the pod, thumbing his fob to reattach the micro-scout and lock the doors as he does so. Seeing what Raphael is doing, he tries his own wrist-link. After a moment, he frowns and shakes his head. He catches up with Celia.

"There's no connection," he says, pulling up his wrist-link to illustrate. "We're on our own out here."

"Finally—a real adventure." She beams with delight. Then she turns to Eiko and her expression softens. "What is it, Eiko? You look unhappy."

"Something is wrong with this place," Eiko says, in a half-whisper. "It's too quiet. Why is no one else here?"

"Because I got our applications fast-tracked," Leo says. "Right to the front of the queue. We're the first."

"Sure you did," Celia says, skeptically.

"It's true." Leo grins. "I know all kinds of tricks."

Celia looks amused, Eiko even more worried.

Just then, Raphael joins them. He holds up his wrist with an expression of alarm.

"We know," Celia says. "No connection."

Raphael looks deflated.

Eiko turns to Ben. "Ben, are you sure you want to do this?" she asks.

"Yes, I'm sure," he says, with a hint of impatience.

"Benito, my man," Leo says and reaches across to pat Ben's shoulder. "We're right here with you."

"Thanks," Ben says, giving him a brief glance.

Finally, a door opens on the near side of the building and two people step outside, a man and a woman. They are a few years older than their visitors, mid to late twenties, and are both wearing plain-looking, dark clothes. The group of five starts moving toward them, Ben and Eiko in front and Raphael bringing up the rear. The man speaks to them when they are still some distance away. His voice is clear and slightly over-enunciated, as if he is unsure whether they will understand him.

"Welcome to transit station West-one."

The group slows, hesitating. The man and the woman stay where they are and continue to smile pleasantly. When the visitors are within arm's reach, the man continues. "Good day to you all. I am Jonathan Parker and this is Catherine Seville." He indicates his companion with a slight movement. Catherine nods and smiles, but keeps her hands clasped throughout the welcome.

"Leo," Leo says and holds out his hand.

"Leo Huang, yes, pleased to meet you," Jonathan says as he shakes the proffered hand.

"Raphael." Raphael has stepped forward on the far side of Celia.

"Yes, of course. Raphael Delacroix." Jonathan tilts his head slightly.

"No—Winter. It's Raphael Winter," Raphael says, looking irritated.

"My apologies," Jonathan offers, with minimal expression. "Welcome, Raphael."

"Eiko Carr," Eiko says with more confidence than she feels. Her hand is half the size of Jonathan's.

"Celia," Celia says, with a smile and the hint of a challenge.

"Charmed," Jonathan responds, but his expression remains unaltered. He turns to Ben. "And you must be Ben Hatton."

"Yes," Ben says, forcing himself to offer a steady hand. Jonathan looks at him for a moment longer than he needs to, it seems.

"Shall we go inside?" Jonathan says. Without waiting for an answer, he moves toward the building and opens the side door. Ben follows him closely, with a determined expression. Eiko keeps up, but is mostly looking at the ground. Celia is scanning the building, alert. Leo is grinning, Raphael scowling. Catherine closes the door behind them.

# 2

## The Day of the Readings

*One month earlier.*

"Ben Hatton?" The councilor said with a quick professional smile.

"That's me," Ben responded as he took the seat across from her. The conversation cubicle was small and impersonal, but well separated from its neighbors by opaque insulators; noise from the large and busy hall retreated. He passed his wrist-link over the sensor and the councilor mumbled something like 'thanks' while moving her fingers rapidly across her desk. From the reflections in her eyes he could see her screen reacting.

"Let us have a look at you." She tapped on her screen and the shared display sprang to life. At the top, he saw his name and underneath it a helix icon with his name repeated. She muttered a puzzled 'hmm' while her face twitched, frowned and finally realigned. She was around thirty, he guessed, so she must have had her own reading ten years ago. Did hers start with a worrying 'hmm' as well? He did not like the sound of it.

"So—all we have is your current sequence file. Not the pre-edit—or any links to your parents' files... That's—well—unusual. And no treatment files."

"I've never been to the hospital or anything."

"But you've seen a family doctor? For checkups and so on?"

"Sure. Aunt Vera."

"*Aunt* Vera?"

"Dr. Vera Weiss. She was our doctor."

"OK." She tapped slowly. "W-e-i-s-s?"

"I think so."

Another 'hmm' escaped her.

"Those other sequence files, do I need them?" he asked.

© Springer Nature Switzerland AG 2020
P. Rørth, *The Unedited*, Science and Fiction, https://doi.org/10.1007/978-3-030-34624-9_2

"Well, maybe not. It depends on your questions. Are your parents with you today? We could just have them link in and then re-attach the files and re-build a most likely pre-edit."

"No. My parents… My parents are gone."

Her eyes rushed to his, in surprise.

"I am so sorry," she said. "Were they…"

"It was an accident," he said. "Two years ago."

"I'm…" She stopped herself. "What are…" She cleared her throat. "What were their names? And birthdates? I'll try to find them."

"Jack and Bella Hatton."

"Previous names?"

"Sorry?"

"Before marriage? I assume one of them had a different surname."

"Oh, yes." He paused. "I don't know."

She glanced at him. Skeptically, he thought. But he really didn't know.

"Birth dates?"

"May fifth—my mother—and September ninth—my father."

"Year?"

"Sorry." He felt her disapproval. "They'd never say." Ashamed of everything he didn't know, he looked away.

"No problem." She said briskly and focused on her screen. "I'll just go for sequence matching." Her fingers slid across her desk again, tapping here and there.

"No luck. I can't find them," she finally said, frowning. "That's very unusual. What…" From her eyes, he saw that new information was popping up on her private screen. For a moment, she seemed rattled. Then she drew a deep breath and settled into another reassuring professional expression before turning to face him. "Let's just go ahead with what we've got, shall we? It could just be the system acting up. We've been having some problems…" She smiled, apologetically. "But first things first," she continued. "If you could place your forefinger right there." She pointed to a small box with a fingertip-sized indentation. "So we can confirm your identity."

He did. "Ouch!" he exclaimed, retracting the finger quickly. It had not actually hurt, but the needle-prick had surprised him.

"Now, let's have a closer look at that file." She clicked on the helix icon and a double screen opened up. One side was the chromosomes, the other a long list of names and numbers. She clicked on one of the chromosomes and expanded one area in a few steps, until they could see the actual sequence—two identical sequences. A third sequence was aligned below, in red. She scrolled along the alignment. Then she did the same for another chromosome.

"Both fine," she said. "Naturally." She folded her hands. "So, do you have any specific questions?"

"Yes, right, I was…" He stalled for a moment, but eventually continued. "Do you see… is there anything I should know about?"

"Well…" She scrolled down through the list on the right-hand side of the screen. "You have good cancer protective alleles, generally. One…" she scrolled on "…no, two of them are not optimal. I've marked them, in case you want to go for somatics." She looked at him directly. "It's borderline. For the tissues where these genes matter, somatics are ninety to ninety-five percent effective." She shook her head. "Sorry, I'm getting ahead of myself. Do you know how somatics work?"

"Sure. I've done premed."

"Of course, yes, it says so right here." She jumped to the bottom part of the screen, where his citizen file was displayed. "So…"

"I'll think about it."

"Right. To apply for somatics, you go right down to the end of this hall." She pointed out the direction. "They'll have you talk to another councilor— and a doctor, a stem cell replacement specialist."

"I'll think about it," he repeated. "Anything else you can see? Anything unusual?"

"I don't see any problematic alleles for diabetes, either, or even for less common diseases. So, no, nothing stands out."

"What about edits?"

"Without the pre-edit sequence, or your parents' files, I can't tell what are edits and what is just… regular genetics." She looked at his face for a moment. He was frowning. "If there is anything specific that you're concerned about," she added, "I can look for it."

He hesitated briefly. "Do I have any of the profiles?"

"Which type? There are so many different traits and combinations. I have to know what to look for."

He hesitated again, this time for longer. Slowly, he leaned in over the shared desk. "Lo-test," he whispered. He pulled back quickly and looked down, at his hands.

The councilor executed several searches, scrolling through each output list as it appeared, before she shook her head. "No. There's no evidence of a lo-test profile being implemented."

"Are you sure? It was twenty—twenty-one years ago. It might be an old one."

"It was twenty-one years ago for everyone here today," she replied, waving a hand about. "So yes, I am quite sure. No profile selected. It's just…"

"Right. Regular genetics." He let it sink in.

"Is there anything else? Any specific health problems in the family?"

He gave her a blank look. "No… Nothing else," he answered after a while.

"Please feel free to contact me again." She touched her wrist-link and held it close to his. His beeped. "If you would like to hear more about the somatic options—or if you have more questions, once you've…" She reached for something attached to a slim box on her desk. "Here's your copy of the file. And the software to read it." She handed him an e-stick.

He took it, but stayed seated, looking confused.

"I'm afraid my next appointment is…" she said, after a while.

"Of course," he said, snapping to and getting up. "I'll let you get on with your day. Thank you for your time."

She looked thoughtful, but responded with a simple "you're welcome." She turned to her private screen and started tapping again.

Ben walked slowly away from the cubicle, lost in thought. Soon, a lively babble of voices reminded him where he was. He looked around. The large hall was full of this year's new adults, many of them excited, or nervous, and speaking too loudly. Some parents had come along, as well. No wonder the place was noisy. He started to feel uncomfortable, hemmed in, but did not know where to go.

"Benito!" he heard someone call. He spun around a few times and finally located the source. Walking toward the familiar face, he pushed away the many confusing thoughts competing for his attention.

"Leo," he said, with a smile of relief. "I thought you were staying away from this circus." Leo seemed to be on his own, as well, but completely calm, unaffected by the crowds. He was not much taller than Ben, but almost twice as broad, all muscle. His upper arms were bare, showing off a couple of simple yet intriguing tattoos. Ben still did not know what the characters meant. Leo's hair was ultra-short and heavily bleached from its natural black. Ben disliked his own hair, which was reddish-brown, matching his abundant freckles, and full of soft curls. But he did not know what to change it to, so he let it be.

"Well, I didn't exactly need a reading, did I?" Leo's expression wavered between nonchalant and bitter. "But I wanted to sign up for my first somatics." He turned and pointed to the rear of the hall. "It's over there. Do you need to get anything done?"

"No, I don't think so. I…"

"You have plenty of time, anyway. There's a three month obligatory waiting period." Leo made a face and shifted his voice to a mocking high pitch. "Just to be completely sure that you are making the right decision."

"Well, I suppose…"

"I've been *completely sure* about this for years," Leo stated, irritably. "Come on. Let's go somewhere. Celebrate—or commiserate." He frowned. "When do you have to show up at that silly job of yours?"

"I start at four a.m."

"Four a.m.! You *are* bonkers." Leo's slap on the back almost knocked Ben over. "But that means we have all night. Tonight's on me, Benito." He looked carefully at Ben's face and added, "You look like you need it."

"I…"

"You don't need to tell me about it. It's the shits for all of us."

Ben could feel himself starting to relax. Leo's company was probably just what he needed at this point. Then he remembered his promise. "I'm supposed to meet up with some old school friends tonight," he said.

"Bring on the friends. The more, the merrier."

"Sure." Ben allowed a smile. 'Safety in numbers,' he thought. Out loud he said: "Let's get out of here."

"Benito! That's the spirit."

\*   \*   \*

"You know, Ben, you look *exactly* the same," Celia said to him, tilting her head first one way, then the other. She was either very drunk or playing it up; he couldn't tell which. Everyone was speaking loudly to be heard over the music. "No older, no wiser, no… Whereas you…" she turned to Leo, who was leaning in over the table. "You look—well, you *are* new, I believe."

"New to you, old to the world. I'm Leo," Leo said, smiling, and sat down in the spot vacated by Raphael.

"Nonsense." Celia frowned. "We're all *exactly* the same age." She flung out her arm to indicate their table as well as the rest of the room. They were sitting in a curved sofa, giving them a good view of the dance-floor and of each other. Eiko was on the other side of Ben, but currently distracted, looking off in another direction. Ben had introduced Leo to Eiko and Raphael early in the evening. Leo had drifted away but returned to the table soon after Celia joined them. "That's why we're here tonight," Celia continued, "celebrating our final step into adulthood: complete self-awareness and self-determination." She sighed, with exaggerated world-weariness. "Well, self-something… I'm Celia, by the way." She looked Leo up and down, appraisingly. He was still wearing the tight-fitting, sleeveless T-shirt that accentuated his physique. "Now tell me, was this selected or did you work for it?" She ran her fingers lightly over his exposed upper arm, stopping short of the first tattoo. Leo was momentarily lost for words.

"Hard work. Hours and hours of it, I bet," Raphael said, coolly. He had come back from the bathroom and was glowering, first at Leo, for being in his seat, and then at Celia, for allowing it. She shrugged. Tall and slender, Raphael towered over both of them. Dislike pulled at his mouth and spite narrowed his eyes. "Your father," Raphael directed at Leo, "or whatever you call him—is not exactly a hulk, is he?"

"Piss off. I don't want to talk about my father."

"Piss off? You're in my seat, buddy."

"Come on, guys," Ben tried.

"Who's your father, then?" Celia asked, turning toward Leo with a sweet smile.

"Can't you tell?" Raphael said. "It's fucking obvious. Victor Huang, the defense contractor guy. The slime ball CEO of Huang Shields."

"Rafi," Ben said. "Leave it, will you? Please?"

"But it *is* obvious."

"It doesn't matter."

"He's a youngtwin," Raphael said to Celia with a big, mocking grin. "He looks just-"

Leo shot up from his seat and pushed Raphael hard, with both hands. Taken by surprise and less-than-optimally coordinated, Raphael fell backwards into the dancing crowd.

"It's kind of funny, actually. Huang senior is so-" Raphael continued from floor level, his voice gradually disappearing in the noise. Someone in the crowd gave him a helping hand and he managed to get up again.

"Don't mind him," Celia said, to Leo. "He's had a tough day. You know, his parents wanted him just they way he was. Isn't that terrible? Wouldn't you be mad if…"

"Cee, don't. That's private." Raphael voice had shed all attempts at humor.

"Of course, Rafi." Celia flashed another saccharine smile. "Does that mean we won't have to hear about how Fran-"

"Celia, please," Eiko interrupted her. She had turned around and was paying attention to the table again. "Let's just have fun tonight."

"Well, then." Celia pulled at Leo's T-shirt, making him sit fully down again. She ignored Raphael's glare. "So, your father…"

"Is an egomaniac."

"And your mother?"

"She's an idiot."

"An idiot?"

"For going along with it. For being an incubator."

"Well, it could be…" Celia tilted her head. "Never mind. I find it terribly interesting." She moved even closer to Leo. "I suspect *you* are interesting, too. You're certainly handsome." She kissed him lightly on the cheek.

Leo seemed confused, but recovered quickly. "And *you* are beautiful." He shook his head as if he had only just realized this. "You're a real doll."

"Hardly," Eiko said, leaning forward. "Do you know what she scored on-"

"Not now, sweetie, not now," Celia half-whispered to Eiko. "Didn't you say 'have fun'?" She added a quick wink and got up from the sofa, keeping her balance easily. "Barbie profile, darlin'," she said to Leo, swinging her long, luscious hair forward and arching her upper body. "Not bad, huh?" She fluttered her extended lashes while keeping her green eyes sharply on his face.

Leo grinned. Raphael looked on with disbelief.

"This doll needs some dancing," Celia continued, pinching Leo's upper arm. He started to get up.

"Sit over here with me, Rafi," Eiko said. "I haven't had a chance to talk to you all evening." He looked reluctant. "Please?" She added. Raphael's hostile pose finally softened. Ben and Eiko slid over to make space on the far side, while trying not to pay attention to Celia.

"You know she doesn't mean anything by it, Rafi," Eiko said, as Raphael sat down next to her. "She's just having a bit of fun."

"At my expense."

"If anything, at Leo's expense," Ben said. "He's an easy target."

"You new friend does seem a bit…" Eiko started. She noticed Ben stiffening and did not finish the thought. "So, how was it today?" She said to Raphael while sliding backwards on the sofa, thereby making it a conversation for all three of them.

"It was fine," Raphael said. Eiko kept looking at his face, but he did not add anything.

"Did François go with you?" she continued.

"No. Why should he?" Raphael responded, slightly aggressively.

"I don't know… It seems a big brother kind of thing to do. He had his reading two years ago, didn't he?"

"Yeah, he did." This had an edge of bitterness. Finally Raphael lightened his tone and added with a shrug: "Anyway, you know what he's like."

"Head in the clouds."

"Exactly. He doesn't notice a thing." Raphael paused. "Not his fault, really."

No one spoke for a while. The music dissolved the remaining tension and the dance floor provided amusing distractions. Celia and Leo were already well into the crowd.

"I didn't learn much," Ben said, looking from Raphael to Eiko. They both looked back. "At the reading, I mean. No profiles, no scary alleles, no…" He shrugged. "The only thing that was odd was…" He stopped talking and picked up his glass. He noticed it had been refilled. Leo seemed to have set something up to keep the drinks coming. Eiko was still looking at him, but he did not continue.

"And you?" Raphael asked Eiko, who gave a small jolt in surprise. "Anything unexpected for you?"

"I…" she responded with a strained smile. "I'll tell you guys some other time. Too much-" she indicated the room, the dancing, possibly the noise, "too much stuff going on tonight."

Ben and Raphael exchanged glances, but did not challenge her. All three focused on the music and the gyrating bodies again. Ben looked back at Eiko and finally noticed the unexpected flash of color. Her hair had always been completely straight, chin-length and pure black. It now had a bright pink streak in it, on the left side. She must have had it done very recently. Not the kind of thing he'd expect from her. He decided not to comment.

"Ah—it's my song!" Eiko suddenly exclaimed. "I have to dance. What do you say?" She placed one hand atop Raphael's and motioned with the other toward the dance floor.

As they left the table for Ben to guard on his own, Eiko sent him an apologetic look. He smiled and mouthed "no problem". Shortly after, he looked at his wrist-link. Almost one o'clock. He might as well stay awake for another couple of hours and go straight to work afterwards. As Leo had predicted he would. Crazy guy, that Leo. Sometimes. But no wonder… Ben remembered what Leo had told him in confidence: that his father wanted another son— with his new wife. This time, however, he wanted a "normal" son. They didn't have the license yet, but any mention of it sent Leo ranting and raving with anger and bitterness. Before that, being a youngtwin hadn't bothered him quite as much. Now he had cut all contact with his father *and* with his mother. But Leo had been a good friend these past two years. Sometimes it was too complicated with old friends, childhood friends. They meant well, but… Ben looked back at the dance floor, spotting them easily. At this distance, tiny Eiko with her tentative, girlish movements could easily be mistaken for a twelve-year-old. Raphael's height added years but his awkwardness gave away his youth. Further away, Celia and Leo were putting on a show. Leo was a good dancer, Ben saw, well attuned to the rhythm and to his partner. Celia was going for the full sexy act, moving herself slowly up and down Leo's thigh, her face all rapture. Ben watched. It was hard not to.

Another dance-favorite started up, the drums and bass pounding away. Now everyone was dancing with exaggerated abandon, as if they all had something that needed exorcising. Maybe they did, he thought. Disappointment? Relief? The readings were over and done with. The time for hoping and guessing was over. From today, they all knew their DNA sequence and—for the most part—how it got to be that way. Now they just had the rest of their—officially adult—lives to deal with.

A girl seemed to be trying to catch his eye. She was standing off to the side, swaying to the music, but not exactly dancing. He looked away, quickly, and kept his eyes from returning. Or he tried to. He did not quite succeed. Perhaps the Dutch courage was working. Perhaps the problem was all in his head. Or perhaps the reading had screwed with his head, making him believe that he could… He looked again. The girl was gone.

\*   \*   \*

A shrill sound penetrated his dream, twisting the narrative. It was an alarm. Fire alarm? There were lots of people, in a big room. Panic rumbled; it spread. The sound continued, mercilessly. He finally reached the surface and immediately lost the dream. The doorbell. It was his doorbell ringing.

The heavy curtains showed bright light along the edges. He looked at his wrist-link. Midday. He had slept only a couple of hours. The bell rang again. Who could it be? No one used that bell. The sound was too damn irritating.

"Ben Hatton?"

"That's me." The deliveryman looked suspicious, so Ben held out his wrist-link for ID. A vague unease kept him from offering a finger imprint. He really hadn't had enough sleep.

"I have a—a package for you."

The man held out a very thin, rectangular item. Ben took it and felt its flimsy lightness. It was an envelope, an old-fashioned paper envelope. Maybe it had an old-fashioned paper letter inside. He was mystified.

Back inside, Ben sat down on the crumpled sofa-bed and turned the envelope over a few times. Why would anyone send him a paper letter? There was no clue on the outside—just his name and today's date. He found a knife and slid it carefully under the flap.

The letter was one page long and typed, but signed by hand. The name was typed underneath, which was helpful, as he could not decipher the signature. "Dr. Vera Weiss." He was surprised that she had used her title—Doctor. He read the letter carefully, frowning toward the end. It wasn't that the text was overly complicated, but, apart from the warm words about how much his parents had meant to her and how much she had enjoyed watching him grow up, he didn't understand what the letter was really about. It was somehow related to his newly acquired adult status, but too vague to make proper sense. He groaned, wishing Aunt Vera were there, so he could just ask her. 'What do you mean?' But she was gone. She had died a few years before his parents' accident. She must have written this beforehand, to be given to him on the day following his reading. She seemed to assume his parents had explained something. But they had not. And now they could not. He frowned again. This was maddening. He went to the kitchen section, put the mug in place and pushed for long black. Maybe a jolt of caffeine would lift the fog.

Sitting on the sofa-bed again, he drank his coffee while staring at the letter and trying to think. He reread the letter. It did not help. Finally, he thought of Eiko. Eiko and her parents. They knew him and they knew his parents and… Maybe they could help. But first, he needed a shower—anything that might help clear his fuzzy head.

*   *   *

"Ben!" Yuriko's eyes lit up. "How good to see you." She smiled at him, affectionately. "It's been ages… Come in, come in."

He stepped inside. She was right; he had not been to the house for several years. It seemed unchanged. The tiny entryway was as neat as always. The ornamental garden out front was a serene display of stones, moss and trimmed greenery. He used to find it strange, a bit dull. Today it had made him smile. He assumed they still had the bigger, less organized garden out back, where he had spent so many hours as a child. Eiko's house had been like a second home to him. Eiko was also an only child and in this they differed from most of their classmates. Her parents, Yuriko and Paul, had always made him feel welcome.

Ben followed Yuriko from the entryway into the main downstairs room. This room looked different, somehow, but he could not tell how. Yuriko was her usual, soft-spoken self. She asked to his summer job—Eiko must have told her something about it—and to his plans for the future. He admitted to the former being not very challenging and to the latter being undetermined.

"It must be very difficult for you," she said, her face filled with delicate empathy. "You must miss them terribly."

"I do," he said, "very much." He was glad that she had not completely avoided the subject, but did not want to dwell on it. "It helps that I live in student housing. There are lots of people around, even in the summer."

"Of course, of course." Yuriko looked away, letting her eyes scan the rear section of the room. She probably expected Eiko to have heard their voices and to have come down by now. "Eiko is in her room," she added, with an air of apology. "Do you remember…?"

"Sure." Ben nodded toward the narrow staircase leading upstairs. The grand piano stood between him and the staircase, unlit and silent. Seeing it, he realized what was different about the room and the house. It was silent. Previously, if he had come by on a Saturday, he would have entered to the sound of Eiko on the piano or Paul on the cello, or both. If not, there would be some classical recording filling the air. Not today. The lid was closed over the piano keys. "I'll go up," he added when Yuriko did not move or say anything further. She looked sad. He wondered briefly if this had to do with his parents, with the quiet of the house or with something entirely different. With a half-nod, she turned away. He mounted the stairs, walked down the short corridor and tapped lightly on Eiko's door.

Eiko looked surprised to see him, almost annoyed. "Ben?"

"Yes. I…" He noticed the splash of pink again. "I like the accent," he said, gesturing loosely to her fringe. "New?"

"Yeah, yesterday. I needed… something." She added a reluctant half-smile.

"Can I come in?"

"Sure." She stepped aside. "Sorry for being a grump."

"No practice today?" he said, tilting his head toward the stairs with the question. "I remember I used to love that: coming here and listening to you play." She frowned. He continued lightly. "Maybe you did more damage last night than I thought."

"I'm just not in the mood. That's all."

Something was surely off with her. Was it recent? He wasn't sure. He hadn't been paying enough attention, he realized. He followed her across the room to the futon. Her room was the same size as his 'mini-apartment', but much cozier. The semi-cluttered table, the fabric on the futon, the wall displays, everything was so Eiko. Even the standard desk screen had the softening touch of a paper flower garland along the edge.

"I've always liked this room."

She looked around, dismissively. "I'll be moving out soon," she said, her voice flat. "As soon as I can find a place."

"But why? It's great here. Your parents are… And it's so close to Uni."

She gave him a look and he did not push it.

"So, what's up?" she asked. Then she frowned, softened her voice and added "you can't have had much sleep." He did not answer immediately, so she continued. "You could get a much better summer job, you know. In the lab or something. I could ask. I mean, herding cleaning bots around an office building… That's not exactly-" she stopped abruptly. "I'm sorry, that was…" She shook her head. He looked away. "Do you think of him when you walk around there by yourself?" She asked softly, tilting her head. She frowned again. "What am I saying? Of course you do. I'm… Just ignore me today, Ben. I'm a bit…"

"It's OK. No one understands why I'm doing it."

"I do. And I should know not to dis it."

He shrugged.

"So, something's up," she said. "Tell me." She stopped. "I'm sorry—again. Manners. Some tea?"

"Sure. Thanks."

Eiko busied herself in the corner of the room while he tried to recollect the pieces of his story. Returning, she handed him a colorful, steaming mug and sat down in the other end of the futon, nimbly folding her legs beneath her. Her expression had in the meantime switched from preoccupied irritation to concern and interest.

"Do you remember Aunt Vera?" He asked.

She furrowed her brow, looking puzzled.

"My Aunt Vera."

Eventually, she shook her head slowly from side to side. "I don't think so."

"I guess you didn't come to our place all that much," he went on. "Dr. Vera Weiss. She was also our family doctor."

"Dr. Vera Weiss." Her face lit up. "*That* Dr. Weiss?"

Now it was his turn to look puzzled. "What do you mean by *that* Dr. Weiss?"

"The famous Dr. Vera Weiss. She crossed over many years ago, had extensive somatics done and went on become a brilliant cancer specialist."

"I know she specialized in cancer genetics," he said. "So I suppose we are taking about the same Dr. Weiss."

"Do you know she donated her cells for research? Tumor cells as well as stromal cells and blood samples. She had samples collected throughout the whole treatment. It's really interesting to compare the changes occurring in the edited and in the…" She stopped. "Sorry. That was insensitive."

"I don't mind. I'm sure she'd be happy to know that what she did… that it was useful. Her work meant a lot to her."

"So she was family? That's a bit…"

"I always called her Aunt Vera. But now that I think about it, I don't know if she was, like, *real* family—and if so, whether she was on my father's side or my mother's side. There wasn't any other family around."

"That's right. I remember thinking you were so lucky. No boring family visits cluttering up your weekends." Eiko made a funny face, but Ben did not see. He seemed to be far away. "I'm puzzled, though," she continued slowly. "If she was…" She paused, glancing her desk screen. "I have an idea," she exclaimed and jumped up. "Why don't we find out?"

"Find out?"

"Whether Aunt Vera is really your aunt. You got your sequence yesterday, didn't you?" She seemed excited.

"Yes."

"Do you have it with you?"

He searched his pockets, pulled out an e-stick, looked at it and handed it over.

"Would you like to find out?" She asked, suddenly hesitant.

"Sure. Why not?"

"I've been working with some of the Weiss cell lines." Eiko sat down by the desk, plugged in the e-stick and started tapping while she talked. "So I have the sequences. We can just run an HLA analysis—or a genome-wide kinship analysis." She looked at Ben's face, as if asking a question.

"Whatever is easiest—I guess I *am* a bit curious now."

"OK. Let me just…" She moved her fingers rapidly over the screen and looked carefully at each window popping up. Finally, she turned back to Ben, shaking her head. "Nope," she said. "She's not closely related to you, that's for sure."

"Family friend, then." He shrugged. "That's OK. I always liked her. She spoke to me like I was a real person, not just some silly kid—like she cared what I thought about things. Plus, she brought me cool presents. Spoiled me, my mother said." He smiled. "She was pretty much the only person who ever came to our house."

They were both silent for a while. Eiko glanced back at the screen.

"So, what made you think of Vera Weiss? Today, I mean."

"Oh, yes, right… I got a letter from her. Special delivery, an hour ago."

"A letter?"

"Yeah, a paper letter. Strange, isn't it? Apparently, Aunt Vera wrote it before she died and somehow made sure it would be delivered to me today, July first. The day after my reading."

"She couldn't have known about the timing, though. Readings were shifted from mid-July to end of June a couple of years ago."

"Right…" Ben mumbled, looking distracted.

"And?" Eiko leaned forward. "What does the letter say?"

"That's the problem. I don't understand it. I mean, I understand the words, but I don't understand the context, the significance, the whatever… She seems to have expected my parents to have told me the first half of the story."

"But they didn't get around to it before…"

"Exactly." He took a deep breath. "Which reminds me—the reading yesterday…" He stopped, took another deep breath and then started again. "You know how I was really angry at my parents before they… before the accident?"

"I'm not sure I do. We sort of drifted apart for a while." She paused. "So things weren't good at home?"

"I blamed them. For years, I blamed them… for how I was. For how I didn't… And it turns out…" The words came out slowly, stumbling, reluctant to be strung together. "It turns out they didn't actually do anything. They didn't select it. It's just… regular genetics." He looked at Eiko briefly, his eyes already filled. "But I blamed them. I said that I hated them for what they had done. And I did. I *did* hate them."

With the final statement, Ben seemed to fall apart. He started shaking and sobbing. Tears were streaming down his cheeks, his eyes no longer focused on anything. This sudden outpouring of emotion was both unexpected and unprecedented. For a moment, Eiko simply looked at him, shocked. Then she unfroze, moved closer and put one of her hands on top of his. She sat like this, silently, until the sobbing stopped.

"Sorry. I'm sorry," he said, wiping his cheeks with his shirtsleeve.

"Don't be, Ben. It's OK. Losing both of them like that… It must be… There must always be things you wished you had said—or hadn't said…" She looked at his face carefully. "But I don't understand. What did you think they had selected?"

He didn't answer immediately. He was still blinking rapidly.

"I thought…" He spoke calmly but his gaze was focused somewhere far beyond her. "I thought they had selected a lo-test edit profile for me."

"Lo-test?"

"You know, to curb aggressiveness and waste of energy, to promote studying and focus."

"I know what lo-test is for… It always seemed a bit—well, overreaching—to me." She frowned. "Whatever made you think they had done that?"

"Just look at me!" He stood up quickly and gestured along his body. She understood. She happened to like that he was boyish in type and not too much taller than her. But, objectively, she could see his point.

"Well…" She shrugged. "Your parents weren't tall either. That's just, you know… They didn't try to compensate. That's all."

"No, that's *not* all." He looked upset and frustrated.

Eiko waited for more. She watched him cautiously. "What do you mean?" she asked, softly, when he had seated himself on the futon again.

Ben looked at her face, but only briefly. His eyes started roaming the room, as if looking for something to focus on. "Never mind, Eiko. Never mind." He sighed. "Sorry." Finally, he looked at her face again. "Anyway, they didn't."

"No. They wouldn't have done that," she said. "I'm sure they wanted you to be… just yourself." She frowned again. "Your mother was a bit strict, wasn't she? Or old-fashioned… And your father…"

"He was always trying to please her. My father was *kind*, you know? He always thought of others first."

"He certainly seemed that way. I'm sure he-"

"My mother used to nag me," Ben continued, not listening, "*all* the time. She went on and on about my homework: was today's done, was yesterday's perfect? She told me again and again how lucky I was to be in a good school and that I should study harder. She wanted me to become a doctor. It was an obsession of hers."

"You could still…" He shot her a quick glance. She stopped.

They sat in silence for a while.

"So, this letter," Eiko finally said. "From Dr. Weiss. Why was it hard to understand?"

"It was mostly sentimental stuff. You know, how we meant a lot to her…" He shook his head. "But there was also this odd bit about a clinic. She gave me the name of a clinic and the name of a doctor there. If I got the letter, it meant that she was no longer around and that I should contact this doctor immediately. She wrote that I could trust this doctor, one hundred percent, that she was discreet."

"Discreet?" Eiko made a disapproving face. "Why would you be concerned about that?"

"I don't know. Honestly."

"Do you have her name? Or the name of the clinic?"

"Sure." He pulled out the letter and handed it to Eiko.

She returned to the desk, typed in a search and tapped through a few screens. "The clinic specializes in somatics. The doctor trained with Dr. Weiss. So there is a connection. What did your reading suggest? Anything that needed doing?"

"You've got my file right there."

"I'm no expert."

"But your parents…"

"Do you want to ask them?" She looked reluctant.

"Not really," he said. "Or at least not yet."

"What did your assigned councilor say? Yesterday. They have to say something."

"She said that there were a couple of alleles I might want to replace. Cancer risk. But also that it wasn't clear-cut. She highlighted them in my file, I think."

Eiko opened his sequence file again and scrolled through the right-hand list. "I see them. From what I know, neither of these are major risk factors. You could get them done, of course." She shrugged. "But it doesn't explain why you would need a special clinic—or a special doctor—a discreet one."

"No."

"That part sounds a bit dodgy, to be honest."

"I agree. But what is it all about, do you think? What didn't my parents tell me?"

"Was there anything else—at your reading? Was there anything out of the ordinary?"

"I'm completely normal. As you see." He gestured toward the screen. "HQV resistance normal, no high risk-" He stopped up. "No, wait, there *was* something unusual. My pre-edit file was missing. Without that, she couldn't actually be sure what was edited and what not, but-"

"Your parents' files?"

"They were also missing, both of them. They were not linked to mine and she couldn't find them—not even when she tried by sequence matching."

"Now, *that* is very strange, indeed." Eiko scrunched up her face. She thought for a while, glancing at her screen a couple of times.

"Do you think…?"

"What?"

"Do you think your parents might have been crossovers, like Vera Weiss?"

"They never mentioned anything like that."

"It could explain your lack of family here *and* the missing sequence files."

"You have Aunt Vera's sequence."

"Yes. But if…"

"If?"

"Maybe your parents just slipped through," she said, tentatively. He still looked confused. She explained further: "Maybe they were illegals and didn't get proper somatic stem cell therapy. It can take years for a legal, unedited crossover to be declared ready. They have to be ninety-five percent for skin, airways, gut… or something like that. And it doesn't always work that well."

"What happens then?"

"They have to stay in quarantine, I think. Or go back? I'm not sure, actually. There are so few crossovers."

"But—Aunt Vera?"

"You said she was their friend. She could have helped them, couldn't she? As the family doctor, she could have kept their secret. And when they wanted to have you, she could have made sure you got the HQV resistance edits and whatever else you needed. It could work. With access to a well-equipped and *discreet* clinic."

"So my parents would have been susceptible?" His voice rose in alarm. "To the virus?"

"Well, yeah… But they would have been safe here. There's no HQV in chartered countries: everyone is resistant."

"I guess, but…"

"But it *is* illegal, *very* illegal."

"They were always very careful, very…" Ben frowned. Reimaging his parents as illegals—their constant worrying and fussing, his naïve irritation and impatience—he couldn't do it. Not right now. It was too much.

"With no edits, they might have had high cancer susceptibility," Eiko continued. "And possibly other problems. They wouldn't have been able to seek any kind of serious medical treatment—not without proper files. Especially after their friend and family doctor was no longer around."

Ben stayed silent. Eiko glanced at him a few times.

"So, maybe," he sighed, "maybe they *did* plan that accident."

"I didn't say that," she answered, too quickly.

"But it makes sense, doesn't it? Maybe one of them was seriously ill, with cancer or something. I don't even know how old they were. They'd never say. They used to joke about it, like it was…" He paused and refocused. "Anyway, if treatment wasn't possible, they might have looked for… a gentler way out. Together."

"But wouldn't they have told you? So you'd understand?"

"Maybe they tried. I was so angry back then. I didn't… I didn't listen to anything. I was such a jerk."

"I'm sure you weren't. You were just confused."

They sat in silence again. Eiko knew they should probably talk to her parents. It might do Ben some good. But she did not feel like bringing it up.

"Eiko!" Ben's eyes were suddenly wide open. "I just realized something. There's another possibility."

"Possibility?"

"My parents." He fixed her with an intense stare and continued excitedly. "I know it's a long shot, but," he took a deep breath, "what if they went back?"

"Back to New Eden?"

"Yes! Their bodies were never found. Remember? What if there *was* no accident? What if the boat was capsized on purpose, to make it look that way? They could have made their way back to New Eden, somehow."

"But why?"

"Maybe someone found out that they were illegal, someone not on Aunt Vera's team. They must have known they'd be put in medical quarantine, or punitive quarantine, for years. They'd have done anything to avoid that, to avoid being separated." He frowned and stopped talking.

"They did seem very close," Eiko added into the silence. "But still—a staged accident—it's a bit… elaborate, isn't it?"

"Or maybe they missed home, the family they left behind. Maybe they needed to see them again, before…" His voice carried less conviction now. He looked directly at Eiko. "No, you are probably right. It *is* crazy. Wishful thinking." He shrugged. "They would have told me, wouldn't they?"

"I think they would. If not in person, then in a letter." She gestured at the piece of paper on the table.

He nodded and stared at the letter for a while. "But—even if we disregard the wishful thinking part—I should have family there, in New Eden—family that I've never met. Grandparents, perhaps. Real aunts and uncles."

"*If* your parents came from there."

"But it does make sense. Given… everything."

"It does."

"I need to find out." He stood up. "I need to go there. Somehow."

She waited for a moment before responding. "Well—as it happens—you might be in luck with that. Timing-wise."

"How so?"

"Some sort of cross-border openness initiative has just been launched. Rafi told me about it the other day. He heard about it from his mother."

"From the minister of science?"

"The only mother he's got, I believe."

"So it's actually possible to go to New Eden?"

"It might be."

"Let's go find Rafi." Ben smiled. "What have we got to lose?"

# 3

## Transit Interviews

"Let me go."

"Come on, Cee. Stay a bit longer," Raphael says, pouting playfully. He has one arm wrapped around her naked waist as she sits on the edge of the narrow bed. The late afternoon sun reaches in, making her skin glow and picking up the shine of her hair. She looks like a goddess, he thinks, too beautiful to be real. He pulls up behind her and kisses the back of her neck. She leans back into his embrace and giggles.

"No, Rafi." She straightens up again. "I need to go back to my room."

"Why?"

"Out of respect for their customs—like a good girl?"

"They think we're damned anyway."

"I know that."

"So why go back?"

"Because." She turns and gives him a generous kiss, then ruffles his hair and pulls quickly out of his grip.

"I'm hungry," he says. He sounds surprised.

"Of course you are." She finds her scattered clothes and puts some of them back on, haphazardly. "You are *always* hungry," she adds as she opens the door. She blows him another kiss and then she is gone.

Raphael thinks about getting up but decides to wait a few more minutes. The bed is comfortable, even if a bit short. He glances at his unlit wrist-link and shakes his head. He listens to the building. It is very quiet. He drifts off.

When Celia enters the dining room an hour later, Ben, Eiko and Leo are already there and are having what appears to be a lighthearted conversation. Like the night before, they have chosen the table by the window. All the other

© Springer Nature Switzerland AG 2020

P. Rørth, *The Unedited*, Science and Fiction, https://doi.org/10.1007/978-3-030-34624-9_3

tables are unoccupied. Why are we the only ones here? Celia wonders, once again. Will they ever explain? The clanging of pots indicates dinner is on its way and the penetrating, burnt smell suggests it will be another local style meal.

"I wonder whether they will serve us old-fashioned meat again," Celia says as she sits down next to Eiko. "From an animal, I mean."

"I had a bite of it last night and had to swallow," Eiko says. "But I hid the rest under a piece of cabbage. I'm afraid the cook will be insulted."

"I'm sure she won't mind," Celia says, with a smile. "I ate all of mine. I liked the taste—and the chewiness."

"I've had it before," Leo says. "Several times. Yesterday was pork." Celia looks unsure. "Pigs. Oink, oink," he continues. Eiko shivers. He laughs.

"We were just talking about our day of leisure—what we've each been doing," Ben says, directed at Celia.

"Benito here was reading a paper book—can you believe it?" Leo says.

"A book is a book." Ben shrugs.

"But there are no links, no clips and you can't search or scroll or anything," Leo says. "It's just … static pages, dead words."

"You're used to reading paper books?" Celia asks Ben. He shrugs again.

"No net access for a whole week—that's just plain crazy," Leo continues. "What are we supposed to do? There's one screen, in a totally crappy room, and of course it only shows flats. A bunch of very old movies, in fact."

"It's disorienting, that's for sure," Celia says. "I suppose it might be a deliberate challenge." She cocks her head. "There's a New Eden news channel, as well."

"Did you watch it?" Leo asks.

"Some."

"They go on and on about the same thing for ages—just talking and talking. Weird. What's the point of watching that?"

"So what *did* you do?" Celia asks.

"I slept. It was great." Leo grins.

"I believe that qualifies as decompression," Ben says, with a quick smile directed at Leo. "So you were doing exactly what you were supposed to be doing."

Leo shrugs. "Not on purpose."

"The grounds are pleasant," Eiko says. "It's like a park, but you have it all to yourself. The forest part is nice, too."

"You didn't leave the perimeter, did you?" This comes from Catherine, who has just joined the table.

"No, of course not," Eiko says quickly. "The grounds are plenty big."

"I guess I should brave the outside tomorrow," Raphael says. He came in behind Catherine and now pulls out a chair across from her, next to Celia. "Today, I was… busy." He looks somewhat disheveled, but content, and smiles pleasantly at Catherine. "I've heard there are some rare trees in this area—some very old ones, too."

"That's correct. That was part of the attraction of this particular location. We have-" Catherine stops speaking as she sees Jonathan approaching.

Jonathan heads for the empty chair at the end of the table. Ben is on one side, Leo on the other. "Excellent," Jonathan says, as he gets seated. "You are all here. I trust your first day was relaxing?" He looks at each of them in turn. Leo looks like he wants to say something sarcastic, but he holds off. Instead, he simply nods. So does everyone else. "Excellent," Jonathan repeats with a brief smile. "Now, you may have noticed that certain things here at the transit center are different from what you are used to. It is more like our daily life in New Eden: simple and here-present, with limited distractions. Catherine and I are happy to answer any questions you might have about our traditions. And our regulations." He waits a moment before he continues. "Now, in terms of your visa applications, the individual transit interviews are quite important. We will start tomorrow morning at nine o'clock." He looks at Leo. "Leo, will you be ready to go first?"

"No problem," Leo says, looking pleased.

"And Raphael, you as well?" Jonathan adds, causing Leo's face to cloud over.

"Right-oh," Raphael says, with a lingering smile.

"We have set up two separate rooms for interviews," Jonathan explains. "One is the room where you had your medical check-up this morning, the other is right next to it."

"And the rest of us?" Celia asks.

"In the afternoon." Jonathan says, adding a minimal nod.

The dinner proceeds much like the first evening. The food is served by the same heavy-set, middle-aged woman pushing a trolley. Catherine thanks her and gets a low, unintelligible reply. As soon as the woman has retreated, Jonathan and Catherine fold their hands and bend their heads. They mumble something in approximate synchrony. The rest of the table waits in uneasy silence, Leo with a slight smirk, the others with as little expression as they can manage. They have read about this old custom, of course, but seeing it for real and up close is quite a different thing. It reminds them that they are truly in a foreign place.

\*   \*   \*

"So, Leo," Jonathan says. "We know who your father is, of course. The founder and CEO of Huang Shields."

"Yeah, I noticed—along the wall. Are those zappers original Huangs or knock-offs?"

"Our defense equipment is all made in New Eden. I'm sure you appreciate why." Jonathan adds a brief, superficial smile. "But I suppose there are similarities to Huang products. Imitation is the sincerest form of flattery, you know."

"Whatever."

"We also know about your father's broader… influence."

Leo groans loudly and almost loses his balance.

It is a few minutes past nine in the morning and just the two of them in the room—plus two hard chairs, one table and two glasses of water. No windows, no distractions. Leo came in first, sat down in the chair facing the door and started tipping it backwards almost immediately. He settled on the optimal angle, hooked a little finger under the table for balance and hovered like this, nonchalant yet tense, while Jonathan got settled and started.

Leo levels the chair with a bang. "I don't want to talk about my father," he says and juts his head forward. "I want to talk about something else."

Jonathan does not flinch. "OK," he says. "Fine." He waits calmly.

"So," Leo starts, looking straight into Jonathan's face. "When we arrived… yesterday, no, the day before, did you detect us coming?"

"We saw you land and we came out to welcome you."

"Before that. With your Sky-Nav."

"Sky-Nav is your system."

"But you hook into it."

"We…"

"I know you do. Just like I know you have access to our citizen files—whether or not we fill in the visa forms."

"I don't know what your father has told you about-"

"Oh, can we *please* forget about my father?" Leo says forcefully. "This has nothing to do with him."

"If you say so." Jonathan says, a bit primly.

"I do say so." Leo mocks Jonathan's tone of voice. "So, hypothetically, *if* you could hook into Sky-Nav via your not-so-secret channel, do you *think* you would have seen us approaching? The five of us and the pod?"

"We might not have."

"Good! So you noticed." Leo's face is brimming with satisfaction. "*I* made us invisible." He stops up, expectantly. But Jonathan does not ask. Leo continues: "I hacked into the citizen safety tracker, disabled the tabs on myself and my friends… and on the pod, of course. I even encoded a space filler on the return signal from the pod."

"Impressive," Jonathan concedes.

"Of course, if we just fell off the grid for days on end, that would be noticed as well. There's a program that… Well, you probably know this already. Anyway, I made fake tabs for us, really good ones, with realistic signals. The algorithms are based on past activities and—get this—if the signal gets tracked, it automatically pings back as low confidence and then resets to a new-"

"Very impressive, Leo," Jonathan interjects. "You're obviously good at this. But why bother? You all applied for visas, so coming out here is hardly unexpected."

"The applications would normally have taken months to go through. Your processing system might be new, but it's slow as molasses and full of holes. It's dead easy to get in and change things, or move them around. I did that as well."

"Hmm." Jonathan looks neither surprised nor agitated.

"You don't believe me?"

"I didn't say… Actually, I *do* believe you."

"The Sky-Nav, the citizen tracker. Those are really tough targets, very well protected."

"As I said, I believe you."

Leo frowns, but does not say anything further.

After a while, Jonathan leans forward and places his forearms on the table. "So," he says, "you wanted to discuss this, discuss your… manipulations, your technical abilities. Why is that?"

"I wanted you to know what I can do."

"Fine."

"And that we are not being tracked."

"Yes." This sounds vaguely like a question.

"Listen, I'm here to help you," Leo says, exasperated.

"Help us?"

Leo looks unsure. For a while, he frowns and fidgets with his inactive wrist-link. Finally, he looks up and continues: "The sample tracking problems at the Reserve. Do you know about that?"

"We've picked up on it. Via our not-so-secret channels." Jonathan nods acknowledgement at Leo. "It seems to be quite a mess."

"I did that as well," Leo says, proudly. "I trashed the whole system. Totally. No pre-em cells, sperm or eggs can be matched to their original files any more. The samples are useless."

"Why useless?"

"No one wants to make babies at random."

"The samples could be re-sequenced and re-filed."

"Well, yes, eventually. But it would take time." Leo smiles. "More importantly, since no one understands what is wrong with the system at the moment, they don't want to touch the actual samples. They don't want to risk losing them."

"They could collect and freeze new samples."

"Then they'd have to make new pre-em panels. All of that would take even more time. What's worse, it would mean admitting that the Reserve has a massive problem—a huge scandal, loss of people's trust, etc., etc." He shakes his head while smiling. "*And…*"

"They still don't know what the problem is."

"Precisely! The whole system needs to be rebuilt from scratch."

"There must be back-up systems."

"Been there, done that."

"Fine, Leo. I'm duly impressed. You are a thorn in the side of the establishment and you obviously know your craft." Jonathan pauses. "But you are not the only hacker out there."

"No, I'm not." Leo smirks. "But I'm the best."

"Hmm."

"It's true. I did this six months ago and it still hasn't been fixed. Obviously, they can't. If they could, they would. So—six whole months with no productive visits to Reserve clinics. You probably know that already." Jonathan acknowledges this with a nod. Leo continues. "Soon, everyone will know. They'll *see* it. They'll *see* the absence of baby-bulges. So far, it's only rumors, contradicted by other rumors, the usual bullshit. But when they finally realize… Everyone will be *so* pissed off." He smiles again.

"Hmm." Jonathan says and strokes his chin slowly.

The apparently disinterested responses and the mock-thoughtful gesture irritate Leo, possibly as intended, but he resists reacting to it. Instead, he leans back, folds his arms over his chest, and waits.

"You said you were here to help us." Jonathan finally says.

"Yes. You guys, New Eden. You hate all this crap, don't you? Everything that is done at the Reserve: the sequencing and choosing of pre-em cells, the edits, the… You want to put an end to that, don't you?"

"Of course we do. It is wrong to manipulate human conception, to engineer and control future generations. It is against God's will. It denies His grace and His wisdom. In fact, we do not accept genetic manipulation of any living being. It is one of our founding principles. As a consequence, we strongly oppose the charter. I'm sure you know all that."

"I do. I also know that what I have done is a huge blow to all chartered countries and their reproductive approach. They are in a panic—or will be, soon. This is a gift for you. It shows how your approach, the natural way, is the right way. It-"

"*All* chartered countries?"

"My worm has infected all the databases, at every Reserve out there. They are all connected." Leo grins. "Idiots."

"But why, Leo?" For the first time, Jonathan looks genuinely puzzled.

"Because... because it's the right thing to do."

"Do you truly believe that?" Jonathan looks unconvinced. "Are you a person of faith?" Leo does not answer. "I know there are believers amongst... in your world," Jonathan continues. "I just hadn't expected... *Is* that why you did it?"

Leo shrugs. He looks around the room, but finds nothing to hold his gaze. Jonathan waits. Finally, Leo takes a deep breath and says: "I did it because I hate him."

"Hate who?"

"My father. Him and everything he represents."

"He's no worse than the others, is he? In the defense business, I mean."

"But he's..." Leo seems exasperated again. "I'm..."

Jonathan fails to suppress a smirk. "You're a clone," he says.

"Fuck you!" Leo jumps up, his chair tumbling backwards. Jonathan flinches from the sudden movement. "Fuck you, asshole," Leo repeats.

"I'm sorry." Jonathan holds up his hands, defensively. "Really, I am." He sounds like he might mean it. "I forgot. I know you don't use-"

"Cloning was stopped eons ago. It never worked well. Clones were defective. They were weak and... Shit! I can't believe you just insulted me like that."

"I'm sorry. I misspoke." Jonathan stands up and takes two steps toward Leo. "Please forgive me. And please sit." He puts Leo's chair back upright and gestures at it. "Please."

Leo sits back down, but defiantly, arms crossed again.

"I apologize for causing offence."

"Have you never met a youngtwin?"

Jonathan sits down as well, delaying his answer. "No, I don't suppose I have."

"There are no monozygotic twins in New Eden?"

"Well, yes, but ..."

"So, what do you do? Do you kill one of them? Both of them?"

"No! Of course not." Jonathan looks aghast. "We'd never do anything like that." He shakes his head. "But identical twins are rare."

"Youngtwins are rare, as well. Anyway, it's the same thing. Twins. Not clones, not monsters, just twins."

"Well, we see this a bit differently. Normal twins are not… deliberate." Jonathan stops for a moment. "So," he starts again. He stops. Leo is slouching in his chair, like a rebellious teenager. Jonathan is shifting back and forth on his. "I've actually met—well, I've said 'Hello' to—to your father—Victor Huang."

"Hurray for you."

"So I *have* met a youngtwin before."

"Technically, he's an oldtwin."

"Right, yes…" Jonathan frowns. "He seems to be a reasonable man—a charming man, even."

"And you are surprised because…?"

"Not what you think. You said you hated him."

"Yes. I do. I hate him. I truly hate him," Leo says the last words slowly, enunciating clearly.

"But why? Why all this hate?"

"What? Are you dense or something?"

Jonathan almost reacts, but stops himself in time. "No. I just want to understand," he says calmly. "So, why?"

Leo says nothing. He maintains his insolent posture.

"If you won't explain, then I'm afraid I can't take this conversation seriously." Jonathan says, sighing audibly. "Just saying 'I hate him' is childish. It is not a reasoned sentiment from an intelligent adult."

They wait.

"What do you think it's like," Leo finally says, angrily, "being predefined like that? Being brought into this world to be as similar as possible to someone else? Even worse, to your selfish prick of father? Not simply to be… yourself?"

"I don't know." Jonathan says, honestly. "I can't imagine."

"Well, it's bloody awful," Leo spits out. "It messes with your… self-esteem and your sense of… whatever. Believe me." He is suddenly calm again. "So, I want to help you stop it."

"Stop it?"

"Yes. Make the world normal again. No more youngtwins. No more… any of that stuff—the picking and choosing."

"And that's it?"

"*That's it?*" Leo says, incredulous. "It's huge."

"So you don't have any message for us from your… from Victor Huang?"

"No, of course I don't! Weren't you listening?" Leo shakes his head. "I want nothing to do with him. I don't talk to him. I left home ages ago."

"We thought maybe the two of you had…"

"Telepathy? We're twins, dammit," he slams his right hand flat on the table, "not one person! Fucking hell…" Jonathan winces. "Sorry," Leo adds. After a while, he frowns and asks "what kind of message?"

"A negotiating position: An offer from your side or a suggestion for how we might proceed—informally, of course. We understand that whatever you bring us would be completely off the record, fully deniable." Jonathan smiles briefly and continues smoothly. "Officially, this visit never happened. Your government knows nothing about it. I understand that part now, clear as day." He nods. Leo looks perplexed. Jonathan observes him for a while. He finally goes on, but his tone has changed. "You don't know what I'm talking about, do you?"

"No. I don't."

Jonathan keeps on looking at Leo, thoughtfully. Finally, in a slow, deliberate voice he says: "So, the only problem you know of at the Reserve is the tracking-and-linking problem—in other words, a computer problem."

"*Only* is not really fair. It's super-sophisticated, like I told you. You can't just undo it, not without… But yeah, that's it… More or less." Leo shrugs. "Of course, they can't trace it to me. I thought I'd better give them an alternative—something logical. So I looped everything via your access route."

"You did *what*?"

"How do you think I know what info you guys look at? I've been surfing in and out on your route for ages. As I said, you really need to do better on…"

"So your hack traces back to our computers?"

"Yeah. Pretty neat, isn't it?" Leo smiles. "I know they've tolerated your unofficial access so far, your occasional semi-serious spying along with the innocent peeking. But if they find out about this—well, they might just change their minds, don't you think?"

"But you… you said you wanted to help us."

"I don't want to get myself in trouble, obviously. And I'm not actually here, remember?" Leo's smile becomes a smirk.

Jonathan sits in silence for a moment, brooding. Then his face lights up, as if something has just occurred to him.

"Did they—did they convince you do this?"

"They?"

"Your government, your father."

"What are you talking about? Of course not! I found your access route by myself. It wasn't all that difficult. I used…"

"Don't bother with the technical stuff. I'm not an expert."

"Right. I got that already."

"I'm talking about your hack of the Reserve. The system is not functioning; the databases are inaccessible. How *terribly* unfortunate..." Jonathan's sarcasm is thick and mocking. "It's just a foil they've set up, isn't it?" He straightens up. "To hide the truth." He shakes his head. "That's... pretty smart, I have to admit. It'll probably work, for a while."

"What will work for a while?"

Jonathan looks at Leo. He looks at him for a long time.

Finally, Leo breaks eye contact. "I don't know what you're talking about," he says. "Fuck. I came here to... to do the right thing. I thought you guys would understand. I thought you'd be thrilled." He throws up his hands. "Blows me. Whatever."

They sit for a while longer, neither of them speaking. Finally Jonathan stands, and holds out his hand.

Leo shakes it, looking uncertain. "So, that's it?"

"Yes, for now. Father Elias will be here tomorrow. He may have some more questions."

"Father Elias?"

"Yes, he's the man your father..." Jonathan sighs. "Never mind."

"So, do I get a pass?"

"A pass?"

"Permission to visit your fair country. That's why we're here, in this room, is it not? The crucial transit interview?"

Jonathan does not answer.

"So?"

"Well, no. We can't let you in."

"You can't *what*? Why not?"

"You're a... a youngtwin. In New Eden, youngtwins are considered an abomination, a mockery of God's will. Some people think you are the devil's spawn." Jonathan says this as if it is, now that he thinks of it, a perfectly reasonable idea. "I'm afraid you wouldn't be safe." He shrugs. "We can't take that chance. I'm sorry."

Leo stares at Jonathan, in disbelief. He has a million things he wants to say, none of them polite. But they get stuck in his head, all jammed together.

Jonathan leaves the room first, with a brief "enjoy the rest of your stay" directed at Leo. He gets no response.

Leo waits for several minutes, alone in the stripped room, apparently frozen. Suddenly, he comes to and storms out. He slams the door behind him as hard as he can. Marching down the corridor, he does not look at anything or for anybody. Seconds later, the front door closes. Much to Leo's irritation, this door is not slam-able.

\*   \*   \*

"There you are," Celia says, as Leo enters the blue room. This small, casually furnished room has already become their designated meeting place. "Where have you been? You missed lunch. The conversation was… interesting." She and Raphael are in the sofa, Ben and Eiko across from them on low stools. Cups are on the low table in between them. Everyone is looking at Leo. "Rafi was just telling us about his little chat with Catherine," Celia continues. She turns to Raphael, but he ignores the hint. "Go on, then," she says.

"It was nothing, really…" Raphael shrugs.

"She was fishing for something, you said?"

"Well… first we talked about my studies, my interest in plants, stuff like that. She knows a lot about… Anyway, that was all fine—just a normal conversation. But then, gradually, it became about my mother: As the minister of science, how was she going to deal with the present situation? What were her plans? What was the government's overall plan? As if my mother would ever tell me that kind of thing." He makes a face—half disbelief, half ridicule. "I didn't even know what she was talking about. What situation? So I asked her." He pauses. "She couldn't—or wouldn't—say. But she kept trying to get something from me. She asked me if I had anything for them to look at. Informally, of course. Had my mother given me a file, a letter, some instructions? It was all nonsense."

"Same thing with me," Leo says. "Jonathan seemed to think I was a messenger from my father. I had no idea what he was talking about."

"Well," Celia says, "I guess we've solved the mystery of how we beat the hordes of ordinary applicants and got to the front of the line."

"No, that was really…" Leo starts, but stops.

"That was what?" Eiko asks, turning around. Leo is still standing and is right behind her.

"Nothing. Never mind," Leo mumbles.

"And why they wanted to talk to you two first," Celia continues, looking at Raphael and apparently not hearing Leo. "Important parents. So, what else is new?" She turns to Eiko. "Eiko, sweetie, you were saying something?"

"Nothing important."

"I wonder what they'll ask us lesser mortals. We're next."

Eiko shrugs.

Leo moves toward a low chair at the far end of the table. He is about to sit down when Ben perks up. "Hetty told me she left some food for you in the kitchen," he says, directed at Leo. "Lunch. It's quite good." Leo sits. "I can go fetch it," Ben adds.

"Thanks, Benito," Leo says, kindly. "But I'm not really hungry."

"Hetty?" Eiko says, at the same time. "Who's Hetty?"

"The cook," Ben says.

"You know the cook?"

"I talked to her for a while, yesterday afternoon. She was making bread. We compared recipes."

Eiko looks at him, still puzzled.

"Good going, Ben," Celia says. "The bread here is fantastic. I'll be coming around to your place when we get back." She gives him a wink. He feels the beginning of a blush.

"Where did you go, anyway?" Raphael asks Leo.

"Out," he answers.

"We looked for you. High and low."

"Out past the perimeter. I needed to get a signal."

"Past the perimeter?" Eiko says, alarmed. "But they'll know. In fact, I think they did know, at lunch."

"I'm sure they guessed. The jamming shield goes quite far, so it was a good jog." He shrugs. "I needed the exercise anyway."

"But if you leave the perimeter," Eiko says, somewhat irritably, "the transit period starts all over. Jonathan said so, when-"

"None of that matters anymore, not for me," Leo says with a touch of drama. They all look at him again. "Our hosts know my father. They also know that I'm a youngtwin. Not that I ever tried to deny it. *But*, to the fair people of New Eden, that makes me an abomination. The devil's spawn or something like that." He scoffs. "Primitive idiots."

"What?"

"He said that rejecting my visa was for my own good. Apparently, I wouldn't be safe in New Eden." He puts on a mock-sad expression.

"But that's not…" Eiko starts.

"So you're not coming?" Ben asks, simultaneously. "I thought…"

"I'm afraid not."

"You're not leaving us, are you?" Celia asks. "We haven't had a chance to get to know you properly." She leans over and puts a hand on his arm, with a playful expression. Raphael draws a quick breath and looks away. "Plus, they might find excuses to reject the lot of us. Your pod would be convenient," she adds in a business-like tone. "I don't suppose hire pods are easy to get all the way out here—even if we could get a proper connection."

"Don't worry. I'll stay for now." Leo smiles. "At least I'm not housebound."

"Thanks." This is from Ben.

"You're welcome, my man."

They fall silent.

Raphael is looking off into space. After a while, he turns to Leo. "Did you learn anything?" he asks, without irritation. "From the net, I mean. Anything that could explain what these guys are getting at?" He turns to the others and continues. "This is not about our visas—it can't be. Something more serious must be going on. And these people think we're involved, somehow, or at least messengers from our parents." He pauses. "Remember, not so long ago, all those unexpectedly appeasing words from the prime minister, directed at New Eden?" He looks around. The others nod. "Then they open this place. Fine, I suppose, it's a goodwill gesture. But it's empty, apart from us. And Catherine's questions... She didn't give me much, obviously, but I did get a sense of importance—of urgency, even."

"I agree." Leo nods at Raphael. "Something's going on. Jonathan's questions were vague, as well, but he kept going at it, or around it, looking for new angles. It might have to do with the Reserve, somehow." Eiko looks at Leo and frowns, but does not say anything. He continues. "They obviously don't mind us knowing that they're fishing for information. That could be significant, as well." He gestures loosely. "Anyway, I did manage to get a signal and connect to the real world. No help. It's just the usual crap. Nothing official, but there's every possible 'fact' or speculation on one news-site or another, endlessly copied and commented upon. I can't tell what's true and what's not."

"One never can," Eiko says with a sigh. "It's so..." She looks up and notices that Jonathan and Catherine have joined them. She falls silent.

Jonathan looks pointedly at Leo. "I'm glad to see you found your way back to us," he says. With a slightly stiff smile, he turns to the others. "So, are we ready for this afternoon's interviews? Eiko, will you go with Catherine? And Celia, if you could come with me?"

Eiko gets up quickly, smiles at Catherine and they move off.

Celia takes her time. As she and Jonathan leave the room, she throws a last glance at the three they are leaving behind. She decides not to worry about them.

"So," Jonathan says when they are halfway down the corridor, "how are you finding the decompression time?"

"Quiet," Celia says and shrugs. "But not bad."

"Have you had a chance to look at our library?" He opens the door to the interview room and gestures for her to take the chair on the far side. Table and chairs been straightened up and the water glasses refreshed. "I think you'll find some books that will interest you."

"I will?" She sits, keeping eye contact.

"Given your interest in finance and economics, I should think so. We have quite a bit on pre-crisis economic models, all the way back to Adam Smith and Maynard Keynes, as well as some game theory texts."

She looks at him coolly, understanding his intent. He returns the look, showing that she is supposed to.

"So, you've checked me out."

"Of course. Celia Saunders. You are one worth watching."

"I'm not an important person, you know. Not like Leo and Raphael."

"You don't really believe that, do you? They have parents in positions of power. It would be unwise to ignore that fact."

"Whereas I…"

"Whereas you will be going places yourself. That is obvious."

"Obvious how?" Since she is fully aware of being buttered up, she feels that stretching it out does no harm.

"Top of your class, even with privileged youngsters like Raphael Delacroix Winter and Eiko Carr as competition," he says with a nod of appreciation, "plus your early entry into bit-market trading. It created quite a stir."

"You know my trading bot?" She asks before thinking. He surprised her.

"I have an interest in such things," he answers, enigmatically. "So will you go academic or commercial?"

"Commercial. I like to wrestle with real stuff."

"You'll be good."

"And rich." She smiles briefly. "OK, enough of that," she continues, her tone no-nonsense and practical. "You've done your bit to flatter my ego, which I appreciate… So, what do you want from me? I'm afraid that I don't know any state secrets. But I *am* curious what this is all about." She smiles again. "Ask away."

Jonathan seems rattled for a moment, but recovers quickly and continues in a more formal tone. "The purpose of these interviews is primarily to help us understand why you wish to visit New Eden and what you hope to learn from the experience."

"Right—so we'll play it that way," she says, looking disappointed. She continues in a lighter tone: "Learn from the experience?"

"As I'm sure you know, this program… policy… of openness is new to us all. We would like to ensure that it is not misused."

"So you think openness might get misused?" She sounds mildly sarcastic, but more like teasing than hostile.

"That the program might," Jonathan says, with some irritation. "You may also know that many people—on both sides—are critical of the new policy and would like to see it fail. We are trying to ensure that our first visitors have an open mind—a genuine interest in our country and a positive attitude toward our traditions and values."

"OK… Genuine interest in a society with a large and poorly regulated barter economy, non-existent resource management, state-controlled foreign trade only, severely restricted personal development, especially for women, and extensive, practically exclusive, political power for a self-selected and by all appearances rather smug and self-righteous religious class?"

"Well," Jonathan says, sliding back in his chair. "You seem to have a complete set of opinions about us already." He takes a deep breath. "And this, naturally, gets me back to the purpose of this chat: Why did you apply for a visa? Why do you want to spend weeks of your life getting to know our world? I'm sure you have plenty more… what shall we say… productive ways of spending your time."

"Curiosity, I suppose." She shrugs. He waits. She sighs and shifts in her seat. The tension abates a bit. "Also, Eiko is my best friend. She and Ben really wanted to go. I suppose you know about that—why Ben is here?"

"And Raphael? He is not your best friend?"

"That's different. He's my boyfriend—on and off." She tilts her head coquettishly, keeping eye contact. He does not react. "Anyway, we all hang out quite a bit. Rafi has been friends with Ben and Eiko practically forever. Leo is Ben's friend—and a more recent acquisition. We don't know him all that well." She pauses. "He seems to be an interesting character, this Leo. Quite intense."

"Yes."

"He thinks you guys are up to something. Rafi does, too."

"And what do you think?"

"I think everyone is up to something."

"I suppose there's some truth to that," he says, adding a brief, seemingly reluctant, smile.

"But you are obviously not going to tell me about it."

He does not answer immediately. She shrugs and looks away. For a while he studies her face—not with admiration, it seems, but with curiosity, as if trying to read her. "I feel I should ask you this…" he finally says, slowly, "given… Well, I'm sure you know that our society has some basic principles, one of which is no genetic modification of any living beings."

"Are you going to bar me from visiting as well? For my own good?"

"No." He twitches. Had he expected Leo not to tell, she wonders. "I would like to hear your opinion," he continues, "or your feelings, about this principle, given that you are…"

"Edited. We all are, directly or indirectly. And you know that. It is one of *our* basic… principles. Well, let me see…" She feigns deep thinking. "I suppose I'm quite happy not to have to worry about HQV infection, or premature cancers, or any number of other preventable diseases. In fact, I find it hard to understand how you justify your resistance when you see people dying, people who-"

"Not everything you are told is true," he says forcefully. "But I'll be honest. It does come at a cost, the fact that we value human dignity and God's grace over physical perfection and increased lifespan."

"Perfection," she scoffs, "what nonsense." She shakes her head. "I won't discuss God with you. *If* such a being exists—which I sincerely doubt—I would not be privy to his or her thoughts about human beings and morality. It seems awfully arrogant of you guys to be so sure..." He is about to say something in reply. She waves him off. "But let's not go there. Or maybe—yes—I'm curious. What about the soul? I thought you guys were all about the eternal soul. And yet, here you are, all sanctimonious about DNA sequences instead. Why does that matter so much? Do you think I don't have a soul because I am edited? Or a defective soul?"

"That's not the point. It's not about-"

"And from where did you get a monopoly on human dignity? How is it that you have it and I don't?"

She stops and looks at him steadily, challenging him.

"We believe that every human being, each with its unique abilities, shortcomings and potential, and each with unpredictable contributions to our shared life on earth, has absolute, unassailable dignity in its own right."

"But apparently not youngtwins."

"That's... Let me finish, please." He clears his throat. "We believe that choosing what future children should look like and be like, designing them, as your people do, robs them of this dignity and thereby of a core aspect of their humanity—an aspect of their soul, if you will. It presumes that you—that we—always know better."

"This not my area, exactly, but some things we *do* know. Some diseases are preventable. We choose to do something about that. Also, some alleles of some genes are correlated with better performance in some areas. So what? Does changing my DNA to include those alleles make me less human?" She pauses. He does not answer. "Saying that we are designed is ridiculous, by the way," she continues. "It's total nonsense."

"Your DNA blueprint was actively selected. One of several fully sequenced pre-em clusters was chosen as the most favorable, then edited further to match some intelligence or beauty profile or whatever." His face is harder now. "That doesn't bother you? Being designed to be how you are?"

"The usual designer-baby bullshit…" She shakes her head. "I should have known that would be your line." She sighs, with intent. "Look: Yes, I have a few edits. I inherited HQV resistance and some of the early cancer-preventive alleles from my parents. On top of that, they altered a few more base pairs in my genome. But it's a genome of three *billion* base pairs." He looks uneasy. "You started it," she adds briskly and continues. "We've just had our readings. We have full access to our genomes, so we know exactly which edits were made. You must know this—since you seem to have a complete set of opinions about us already."

"It's not the number of changes, it's the principle of-"

"And if you think you can design a human being that way, you don't know anything about biology. Even *I* know that much."

"But still," he says, visibly irritated, "for your parents to decide all this beforehand, for them to tweak and modify you, not being content with who you might be if you were simply… yourself. You're saying that doesn't bother you?"

"I *am* myself, you moron!" She glares at him. "I am very much myself." She shakes her head in disbelief and starts getting up from her chair. "And this particular self has had enough of this ridiculous conversation."

He stays seated, his lips pursed, and does not respond.

As Celia is leaving the room, she stops in the open doorway. "You really should think about a career in diplomacy," she says loudly, her voice heavy with sarcasm. Then she shuts the door—a little too forcefully.

In the next room, Catherine smiles apologetically. "Jonathan can be a bit harsh. It tends to provoke people."

"I'm not worried about Celia," Eiko says, with a half-smile. "She's no push-over. Let's just continue."

"Right." Catherine glances at her notes. "I didn't actually expect you to know the details of your parents' work at the Reserve. But I was told to ask you. I hope you don't mind."

"Not at all. It's fine."

"Good. So, let's talk about you and your reasons for being here. You mentioned that you wanted to help your friend Ben?"

"Yes. He's my oldest friend. We used to…" She falters.

"In his application, Ben says that he thinks he may have family in New Eden. Has he discussed this with you?"

"Yes. His parents…" She stops and frowns. "Maybe he should be telling you this, not me."

"Don't worry. He's the next person to be interviewed. But it seemed sensible to get some additional perspective from the person who knows him the best. That would be you, don't you think?" She smiles reassuringly.

"I guess so. We sort of grew up together."

"And you were saying—about his parents?"

"Jack and Bella Hatton. They passed away a couple of years ago. An accident."

"I'm sorry to-"

"Actually, I shouldn't say that," Eiko adds quickly. "They are presumed dead. It was a boating accident. As far as I know, they... well, their bodies were never recovered." She shakes her head. "I don't know the details. You should ask Ben."

"Of course. So, was it hard on Ben, losing his parents so suddenly?"

"He didn't talk about it much. But yes, I'm sure it was." Catherine nods a few times and Eiko continues. "After a while, the authorities told him they were convinced his parents were dead. He seemed to accept it at the time. But recently he's started thinking they might *not* be."

"And other family?"

"That's part of the problem. It's always just been him and his parents and... no other relatives. So it's..." Her eyes roam, sliding across the bare table. "I know he wants to find out as much as he can about his parents. And he'd like to meet his extended family, if possible." Eiko looks directly at Catherine again. "I think it would be good for him. To get some... answers."

"Family is important, I agree." Catherine smiles gently and pushes some stray hair behind her ear. Her eyes are kind, Eiko thinks. Catherine looks at her notes again before she continues. "But why, exactly, does he think he has family in New Eden?"

They talk about Ben's parents for a little while longer. Eiko has trouble recalling any specific details about them, so her descriptions are short and vague. She explains about them being relatively isolated and having limited interactions with the authorities. This is based on what Ben told her recently. She remembers the missing sequence files and mentions that, as well. For some reason, not quite clear to herself, she does not mention Aunt Vera. The argument for Ben being connected to New Eden therefore ends up being rather weak. This does not seem to bother Catherine, who remains interested and sympathetic throughout. She reiterates that they will talk to Ben about his family and compliments Eiko on being a good friend. They move on to talk about other things. Eiko asks questions about New Eden and their belief system. Although planning a career in science, she has an interest in the spiritual side of things, she says. Catherine explains. Neither of them seems eager to break up the conversation.

Eventually, Eiko says goodbye to Catherine and leaves the room. On her way down the corridor, she starts rehearsing what she will say to the others, and, remembering Celia's early and more dramatic exit from her interview, wonders what she will have said. But there is no one in the blue room. Eiko stands in the doorway, looking at the empty sofa, the misaligned low stools and the orphaned coffee cups. She suddenly feels very lonely. Irrationally so, she knows. She shakes it off and heads for her room.

*   *   *

Celia is walking fast, as fast as the forest will allow. She passed a sign that indicated the perimeter, officially called the end of the transit zone, some time ago. The path more or less ended there. It was just a small sign. No fence, no barbed wire, just a reminder. Almost like a dare. She knows that running off like this is not doing any good. She knows she should have waited for Eiko to come out. But she was too angry to stay. She is still angry. The argument started almost immediately. Raphael was standing in the doorway to the blue room, apparently waiting for her, when she came striding out. Leo and Ben were no longer there.

"What happened?" he asked.

"That arrogant bastard. First he tries to flatter me. Then he insults me."

"Insults you?"

"The kind of crap you'd expect from these people: How does it feel to be a designer-baby? Edited to perfection? Not truly 'yourself'? Naturally, *we* are the guardians of human dignity and of the human soul. We're not entirely sure you have one."

"That's just…"

"I explained. We are talking about a few changes to my DNA. How does that make me not myself? Less 'true', less of a human being? What kind of bullshit is that?" She slowed down. "He was probably just trying to wind me up."

"I'm sure he was." Raphael paused. "You know, they must see you as…"

"As what?" She snapped.

"As…" He hesitated, but then jumped right in. He should have known better. "You've called it the Barbie profile yourself."

"As a *joke*. And on myself." She looked at him with fury. "Anyway, that wasn't what my parents went for," she added, after a moment.

"Whatever they chose, it worked like a charm. Fantastic test scores, across the board. You have to admit…"

"Don't be a jerk, Rafi. Do you think I didn't work for my exam results? Do you think it was handed to me?"

"I didn't say that. You studied hard. I know that." He held up a hand, defensively. "But perhaps, just perhaps, the choices your parents made some twenty-one years ago also had an impact. They pushed you ahead."

"Pushed me ahead of others, you mean?" Her eyes were hard, direct. "Unfairly?"

"Well," he said, looking down, "they did everything they could for you, didn't they?"

"*My* parents did?" she said, angrily, incredulously. "As opposed to yours? What the fuck! You've got fantastic parents," she continued, despairingly. "They're intelligent, well off, caring…" She looked at him more closely. "Is this about your brother again?"

"No. We were-"

"I am *so* tired of hearing about François."

"But I didn't say-"

"Self-pity is not appealing, Rafi. It's pathetic."

"It's *not* self-pity." Raphael's face was starting to color. "It's about what is fair and what is not fair."

"Rafi, don't be an idiot. You are smart. You are handsome. You are pretty much perfect." She threw up her hands, in defeat.

"I didn't…" Raphael stopped and exhaled, loudly. "We were talking about you—you and your little chat with Jonathan—why he upset you." He stopped again, this time for longer. His expression shifted slowly into a smirk. "Maybe he found a sore spot, huh? Did he grill you about your edits? You still haven't told me what you got. What are you hiding?"

"Nothing," she threw back. "But my file is my file. My DNA is my DNA. And *if* I wanted to talk about it, I certainly wouldn't pick someone as obsessed with this whole business as you are. Ever since the reading, you've been moody and irritable and no fun whatsoever. Your parents this, François that… It's unhealthy."

"I'm just being honest. Whereas *you* are being defensive."

"Fuck you. *You* are being are *offensive.*"

"Well, fuck you too."

She does not remember which door she slammed. She does not remember leaving the center. She remembers walking down the path, rapidly. She remembers not hearing any footsteps behind her. Raphael did not follow her. In anger, she walked even faster. The forest finally closed around her.

She will have to turn around soon. She has to go back. She knows that.

# 4

## The Truth Will Set You Free

At breakfast, scraping of cutlery is heard, slurping of coffee, but no one is talking. Celia comes in last, collects some food and sits down. Eiko looks up, trying to catch her eye. She does not succeed. Raphael is staring very hard at his cup. Ben is off in his own world, as is Leo. Jonathan and Catherine are absent.

"It seems we all missed dinner last night," Eiko says, looking around at the bent heads. "Catherine told me this morning." No one responds. "She's actually quite nice," Eiko continues stubbornly. "I think she wants to help."

"Yeah, she's OK," Raphael says and sends Eiko a feeble smile.

"Jonathan is an ass," Leo mumbles, but without much emotion.

Celia looks up and her expression suggests she is about to say something harsh. She stops herself and focuses on her food again.

Eiko looks at Celia, then at Raphael, inquiringly. He makes a 'who knows?' gesture before he reengages with his cup. Finally, Eiko turns to Ben and asks "do you know who is going to interview you?"

"Someone called Father Elias," Ben says. They all look at him, even Celia. "He's coming in this morning, apparently."

"Who is Father Elias?" Raphael asks.

"I don't know," Ben says. "Jonathan just told me the name. He didn't explain."

"He's…" Leo starts, but stops again, frowning.

"I told Catherine about your parents," Eiko says quietly, to Ben. "She asked me about… you know, why we had come and… I hope you don't mind."

"Not at all," Ben says. "I wrote about them in my application."

"I didn't tell her anything about Aunt Vera," Eiko adds.

"Good," Ben responds with a quick nod. "At least I think so."

© Springer Nature Switzerland AG 2020
P. Rørth, *The Unedited*, Science and Fiction, https://doi.org/10.1007/978-3-030-34624-9_4

Raphael looks at them and asks: "Who is Aunt Vera?"

Eiko is saved from answering by Catherine arriving at the door. "Good morning everyone," she says, cheerfully. Their replies are subdued. "Ben," she looks at him and smiles, "will you be ready to start in half an hour?"

"No problem." He sounds eager, but looks more apprehensive. "I'll be there."

"Great. Room one." Then she is gone.

Ben pushes back his chair, but remains seated for a while with his eyes on the empty doorway. Eiko looks at his face. She imagines she can see slivers of fear running through the innocent hope. Gradually, his expression becomes more determined. He gets up and walks off with a quick "later, everyone". The table falls quiet again.

*   *   *

Father Elias is older than Jonathan and Catherine, for sure. But precisely how much older Ben cannot say. His hair and his beard are gray, and this dominates the initial impression. He wears a dark frock, which falls straight to the ground and hides the body completely. His face is long and heavily lined, but strong, somehow, the eyebrows unexpectedly dark and bushy, the eyes narrow and still. His voice is deep. Ben concentrates on his words.

"Good morning, Ben. I am Father Elias. I am glad to meet you."

"Thank you. Likewise." Ben nods once, in greeting, as no hand is extended. They sit down across from each other.

"So, Ben," Father Elias says, looking calmly but intently at Ben's face. "First of all, let me assure you that I am here to help you. I wish to help you." He waits a moment before he continues. "But, for that to work, the two of us must trust one another. We must be honest with one another." Ben nods again but does not really understand. "Will you trust me? Even though we have only just met?"

"Sure." Ben clears his throat. "Yes, Father."

"Now, from your visa application, I understand that you have a special interest in New Eden. You think you may have family here."

"Yes… I believe my parents, Jack and Bella Hatton, may have been born in New Eden. I'm not sure, but… I'd very much like to find out. And, if I have extended family, I'd like to meet them. I'd like to know more about… my background."

"I understand. And we will discuss that. But," Father Elias pauses and inserts an enigmatic smile, "you came here as part of a group."

"They are my friends. They came along because they wanted to help." Ben thinks for a moment. "I suppose they also thought it might be a cool sort of adventure... doing something none of their other friends have done."

"So you know them well?"

"Sure. I've known Eiko and Raphael forever. And Celia—well, for years. The four of us used to be in the same class at school. We still hang out. Leo, he... I know Leo can be a bit... much. But he's actually a really good guy."

"Right, right." Father Elias looks down, contemplating something. Ben waits. Finally Father Elias raises his head again, looks Ben in the eyes and smiles. "So, Ben, tell me about your parents."

Ben does. He describes what they looked like, their jobs, their hobbies, their secluded home life and the absence of extended family. He mentions their DNA files being inexplicably absent from Med-data. Like Eiko, he does not mention Aunt Vera. He has been thinking about her since he got the letter and now remembers her disparaging remarks about New Eden—the country in general and the self-important priests in particular. Some of this was voiced in private, to his parents, but some of it was said in public, he thinks. She cannot be a popular person here. Ben finally describes his parents' accident and the presumed outcome. He admits he has no evidence that they are still alive, but he does not want to give up hope. Father Elias nods at the mention of hope, his face set with what appears to be kindness. He does not ask any further questions, nor does he volunteer any answers. Ben soon runs dry.

"These past two years must have been very difficult for you," Father Elias says, after a lengthy silence.

"Yes. It's been... scary, a bit—and lonely, sometimes. Often." Ben stops, realizing it may be the first time he has said this out loud. "But..." He frowns. "Now I'm mostly confused. Everything is so... I don't know anything, you see. I never knew why I had no other family. I never asked. They never told me where they came from, or that they were... unedited."

"Pure. We prefer the term pure. We are humble but pure expressions of God's intent." Father Elias smiles briefly. "Genetic manipulation by selection and editing removes human beings from this sacred state. We believe that each human being is unique and special—from his miraculous inception until his natural end. He must be accepted—cherished, even—for who he is. Pure." He smiles again, a faint but lingering smile.

"That's nice." Ben pauses. "The idea, I mean. I get it." He pauses again. "Of course, I'm just... normal." He looks briefly at Father Elias' face, then down. "I have the required edits, that's all... I think... They had to, you understand... But I fully respect..." He looks back up, pleading.

"But, Ben…" Father Elias looks at him with surprise. "You are *not* edited. Not at all." Ben's expression mirrors the surprise. "You are like us, like your parents," Father Elias continues. "You are pure." He pauses. "You didn't know?"

"No, I…" Ben stops, looking very confused. He thinks for a while. "I had my reading last month. The councilor checked and said everything was normal. The HQV edits were there. Otherwise they would have…"

"You do not have the HQV edits."

"But I do. I have seen my file. I don't know why-"

"You had a medical when you arrived here. Do you remember? You gave a small sample of blood. I am told the sequencing was clear as day. So, we are quite sure. You *are* pure, Ben."

"But… what about my reading?"

"I don't know," Father Elias says, shaking his head slightly. "Your file must have been tampered with. By someone with access."

"But who…" Ben starts but does not finish. "That's crazy," he says, instead. Father Elias does not comment.

Ben stares hard at his own hands, resting in his lap. His mind is reeling, trying to work everything out. "You know this," he finally says. "*You* know that I am… unedited. But the authorities, Med-data, they still think…"

"I'm afraid not." Father Elias sighs. "They know now."

"You told them?"

"Of course not. We would never do that." Father Elias pauses, his expression firm. "But, as of two weeks ago, you are in their system as unedited. Unsurprisingly, you are also flagged for urgent follow-up."

"But how?" Ben finds his own answer, remembering the councilor's harried look, her evasive answers and the unexpected finger-prick. "Anyway, I don't want—I don't need follow-up. I'm an adult now. I get to decide."

"Not about that, you don't." Father Elias shakes his head, his voice full of sympathy. "The rules are quite clear."

"But I…"

"Your travel privileges have also been suspended. You are not to leave chartered territories. They have let our transit centers know. My guess is they intend to quarantine you."

"So I'm screw… I'm stuck." Ben looks panicked. "You're saying I can't go to New Eden. I came all this way. I get so close and now I can't… And if I go back they'll…" He turns his face away.

Father Elias remains quiet and still, his hands clasped and covered by the long, loose sleeves. "It doesn't have to be like that, Ben," he says, after a while. He waits until he has Ben's full attention before he continues. "They've sent an alert, which we may or may not have seen. It is also on your file, which means we can't process you officially. But, the thing is, you might not even be here." He stops, catches the puzzled look from Ben, and continues. "They don't know where you are."

"The citizens' safety people can tell them where I am. We're all tracked." Ben holds up his wrist-link as illustration.

"I know that. But apparently," Father Elias pauses and flashes a wry smile, "your friend Leo Huang has somehow managed to conceal your group's movements. Or so he told Jonathan."

"Really?" Ben says, excitedly. "Bloody hell, Leo…" He looks over, worried about the swearing. Father Elias does not seem to have noticed.

"He has you walking around in your regular life. We've checked this, discreetly. It seems to be working."

"Wow! Cool."

"This does give you some freedom, at least for a while." Father Elias does not explain further. Ben understands and nods slowly.

A few moments later, Father Elias leans back in his chair and clears his throat. "So, Ben. I don't know how familiar you are with New Eden—its history, its culture, our daily life."

"Not very."

"I assume you were taught about the crisis in school? You were told why there are chartered territories and independent territories like New Eden? You know why the wall was built?"

"Of course. We get all that in primary school. But that's ancient history, isn't it?"

"Much of it, yes. But it is still very much with us." He speaks calmly and seriously. "It was essentially a clash of religion and science. It started with the clones and protests against the clones. It ended with forced separations. People were driven from their ancestral land."

Ben shrugs but, shortly after, frowns deeply. "What about the virus? I thought HQV was the real reason—for the crisis, for the wall."

"The real reason… You know how history is written, don't you?" Ben nods, but only just. "The history you were taught is the chartered territories' version. We have a different perspective. Our view is that the insults against human dignity *had* to be stopped: first the clones, then the mandatory manipulation of unborn children for fear of disease. Our forefathers felt that they had a duty to stand up against it. And they did, risking their lives for their principles. Their intent was honest and good. They held peaceful demonstrations. But the government would not allow dissent. Unfortunately, a few extremists got impatient and some events got out of hand. That gave the crisis government the excuse it needed. They established the charter and built the wall. Our forefathers were relocated by force." Father Elias sighs and looks down. "I won't insult your intelligence by claiming they did no wrong. Mistakes were made, on both sides." He pauses. After a while, he continues, looking Ben straight in the face again. His expression is gentle. "You will have to decide for yourself who you are and where you belong, Ben. In New Eden, we believe in something higher than ourselves. This gives us dignity and strength, purpose and belonging. It is hard for me to explain this fully. It is something one must experience for oneself, I think."

Ben nods. He looks thoughtful. "And what about HQV? We've been told so much about it. Are you saying the virus doesn't exist?"

A trace of annoyance crosses Father Elias' face. "A virus should not rule the world," he says, his voice firm. "The HQV virus *did* exist. It *did* claim lives. But the human toll was greatly exaggerated and the fear of infection deliberately stoked. It was straightforward, cynical scare tactics. They made people believe they could choose life over death by complying with the charter. A cunning and effective approach, but it had far-reaching consequences. We are now strictly divided—by a wall, by… biology and by how we see the world." He softens. "It seems we cannot reach across this divide. This is sad." He looks sad.

They are silent for a while.

"Unedited," Ben muses. "I would be an illegal, now."

"Exactly. If you go back, you will be kept away from us—from your family, from knowing about yourself and your heritage—forever."

Ben nods and straightens up in his chair.

Father Elias starts talking again, softly and evenly. "What I would like to do with the rest of our time, Ben, is to tell you more about New Eden. I will tell you what life is like with us—from the perspective of an old man who has spent a long and contented life here." He smiles. Ben returns it. "I hope this will help you understand—and help you decide what to do." He pauses. "I will tell you about our faith and the humility it teaches us, about our communities, our respect for labor and for the land, and the importance of family." He pauses again, looking searchingly at Ben. "If you would like me to, that is."

"Yes," Ben says, with unexpected force. "I would like that, very much."

\*    \*    \*

"So?" Eiko is with him immediately. She must have been watching the door. "What did he say? Does he know where your family is?"

"He…" Ben starts, his voice cracking. He looks startled, like an animal caught in headlights. "No, he… Can we go outside? I need some fresh air…"

"Sure." She starts to turn around.

Ben hesitates. "In a minute, maybe?"

"Of course. I'm sorry for rushing you. Out front in five?"

"Sounds good."

He arrives a couple of minutes late, looking calmer. They take the winding path around the grounds. They pass freshly cut areas of grass, a couple of well-tended flowerbeds, some bushes. Finally Ben breaks the silence.

"I'm one of them, Eiko."

"What do you mean?" She frowns.

"I'm… unedited."

"What?"

"My parents must have been, like we guessed. But so am I, Eiko. So am I."

"But…" The frown turns to disbelief.

"They checked," he continues, "as part of the medical we had two days ago."

"No…" Eiko stops walking and Ben does the same. She looks him straight in the eyes. "No, Ben, that can't be right. You just had your reading, remember? It showed you had… at least the required edits."

"Father Elias says someone must have modified my file. Aunt Vera, I suppose, to protect my parents. All three of us were illegal."

"And you believe him? You trust the say-so of some New Eden priest rather than the Med-data file and your appointed councilor? That's crazy."

"He says that I'm flagged as unedited in Med-data as well—as of two weeks ago. I'm not allowed to travel."

"You can't be…"

"It makes perfect sense, though. My parents' disappearance and the missing DNA files—yes—that could be due to the two of them being unedited, illegal and suddenly scared or very sick or something. But the letter I got from Aunt Vera, that was about *me*. She urged me to go to a specific clinic to get somatics done. She emphasized the need for discretion. That part only makes sense if I'm unedited, as well."

"Yes, but…" She looks away and purses her lips. He waits for her to work through it. She looks back at him, her face set with determination. "OK, Ben, listen: If you are even half-way convinced that this is true, we absolutely have to go back. Now. We need to get you checked out, get the somatics started, everything. Maybe that clinic will still…"

"But that could take years."

"Maybe. But you don't have a choice. I don't know what your parents were thinking…"

"I won't be able to visit New Eden, ever."

"Probably not. It would be too dangerous. Somatics aren't one hundred percent and the brain is not—you know… But overall, it's quite effective. HQV enters-"

"Father Elias says the whole virus thing was exaggerated," he interjects. "Scare tactics, to make people accept the rules of the charter. And it was long ago."

"Father Elias says…" She scoffs. "You don't believe that, do you?" Frustration adds an unpleasant edge to her voice. "Don't fall for it. Please, Ben."

Ben starts walking again. Eiko hurries to catch up.

"They don't know you," she continues. "They don't care about you. Maybe they are trying to manipulate you for some reason. They are fanatics."

"Father Elias is not a fanatic. We talked for a long time. I trust him."

"Based on one conversation?"

"Yes." Ben stops and turns to face Eiko, a sudden eagerness displacing the residual confusion in his face. "He explained what life is like in New Eden. He talked about faith a little bit, but mostly about purpose, work, nature and family—and how best to contribute to a community. In New Eden, everyone does that. And everyone is accepted for who they are. They are not pushed to fit into a predefined mold. I like that. It feels right to me. And I *am* one of them."

"You can't say that, Ben. You don't know anything about them. Not really."

Eiko fights back tears. Ben does not see. He has started walking again.

"This Father Elias, he can claim anything he wants," Eiko says when she catches up with him. "You don't know what the truth is. You don't know what it's really like there."

"Why would he lie to me?"

"Because…"

"Eiko, I need to go. I need to find out about my family."

"Did he claim that your parents are still alive? Did he say that they are safely in New Eden, just waiting for you to show up? Come on, Ben." She puts a hand on his sleeve. "Please. Let's go back home."

"Home?" Ben stops up and turns around. Eiko retracts her hand. "And no, he didn't talk about my parents," he adds. "I guess he didn't know."

"Please, Ben…"

They stand for a moment on the path, facing each other. Then Ben turns to resume the walk. Eiko follows.

The rest of the loop is completed in silence. As they approach the main building, they see Catherine out front, waving at them. They walk over.

"Lunch," Catherine says, smiling. "Also, Father Elias would like to talk to all of you afterwards."

"All of us together?" Eiko asks.

"Yes—in the blue room. He wants it to be informal. I'm not sure what time it'll be, but…" She turns toward the building. Eiko and Ben follow her.

"Ben," Eiko whispers, somewhat urgently, "let's just go home. Now. We'll find Leo, explain the whole thing to him, and go."

"No," Ben answers, not whispering. "We should see what Father Elias has to say." He sends her a conciliatory smile. "We can decide afterwards, OK?"

"OK," she responds, miserably.

*   *   *

They do not get started with Father Elias until four in the afternoon. Lunch was subdued, Ben fending off a few questions at first and Eiko quiet throughout. They split up soon afterwards, some inside, some outside, battling their mounting restlessness each in their own way. At two, Father Elias started walking the grounds and the corridors, knocking gently on doors, introducing himself with a few words and a quiet authority before saying when they should meet up. No one refused.

"Please, everyone," he says as he enters the blue room, "have a seat, wherever you want." He pulls up an extra chair, allowing the guests their usual seats. Eiko forgoes her low stool and slides in next to Celia on the sofa. Tea and coffee are on the table. He gestures for them to help themselves. They do. They mumble amongst themselves about cups, sugar and spoons. Then they turn to him, expectantly. Leo's curiosity is mixed with hostility, Eiko's with apprehension.

"My name is Father Elias." He sits very still and upright in his chair, the long frock remaining unruffled, but moves his head and smiles faintly as he speaks. "Eiko, Ben, Leo, Celia, Raphael. Welcome." He nods at each of them, in turn. "I thought it would be useful for me to shed some light on the situation here at the transit center—your situation, that is. As I told Ben earlier today, I prefer openness and honesty in all things." He looks briefly at Ben. "It is the best way forward. Don't you agree?" He looks at the others, one by one. They mumble their replies. Agreeing, of course. "Excellent." Father Elias' expression does not change. "So, you have expressed an interest in visiting New Eden and you have applied for visas. You have also, individually, discussed your motivations with Jonathan or Catherine." He looks around again, finishing with Ben. "Or me." Nods from everyone. Some feet shuffle.

"Who are they, anyway?" Raphael asks, a bit too loudly. "Jonathan and Catherine? Are they priests?" He looks surprised to have spoken.

Father Elias flashes a slightly annoyed look, but he answers calmly. "No, not yet. They are in training."

"Here? Interviewing tourists?" Leo jumps in. His eyes are wide open, as if he is forcing himself to mount the challenge.

"Shall we focus on why you are here?" Father Elias asks in place of an answer. No one responds. "Of course, none of you are officially here. You are actually back home—or lost in cyber-space." They all look at him, Leo and Ben neutrally, the others with bewilderment. "Thanks to Leo's impressive computer skills," he adds. All eyes turn to Leo.

"That's me." Leo smiles. "The citizens' safety folks think we are all happily walking around in our usual routines. I had a bit of fun with their tracking system. I even put in noise shuffles and redirects in case anyone gets too curious about us."

"But what if…" Eiko says, in alarm. "My parents. I didn't tell them where we were going."

"So?" Leo shrugs.

"How did you manage that?" Celia asks, looking grudgingly impressed. "Getting into the CS computers must be-"

"More to the point: Why?" Raphael interrupts, his voice raised.

"I also made sure we jumped the queue for this place," Leo says, but then turns to Father Elias and adds "unless you guys engineered that. For your own reasons." Father Elias does not respond.

Raphael slides forward in his seat and continues to stare at Leo. "Tell us why, you arrogant prick," he demands. "You can't just erase us like that."

"Come on, Rafi," Leo teases. "Where's your sense of adventure?"

"You bastard…" Raphael says with an abrupt, rude gesture. Celia pulls him back.

"In any case," Father Elias continues in a normal, pleasant tone, "the situation is actually quite straightforward now." He looks around. "Isn't it?" The confusion in their faces belies his statement. He continues. "While we appreciate your interest, we cannot go forward with any of your applications."

"Why not?" Raphael says. "We've done everything…" He stops and looks away.

"Three of you obviously aren't serious about visiting New Eden. When welcoming you here, we specifically asked that you stay within the perimeter and off the net. We also told you why. You chose not to respect our request."

Celia looks over at Eiko, lifts her hands in apology, and mouths "sorry". Raphael is still looking away and hopes no one will ask when and why he ran off. No one does.

Father Elias keeps his eyes on Leo. "Our tracking system is different," he explains. "I guess you didn't find it."

Leo shrugs. "Jonathan told me I wouldn't be allowed in because I'm a youngtwin. So I had no reason to care about your requests and regulations."

"Yes." Father Elias frowns. He turns to Eiko. With what sounds like sincere regret, he addresses her directly. "I *am* sorry. But those are the rules. We do not allow youngtwins in New Eden. We cannot guarantee your safety."

All eyes turn to Eiko, who looks down. An abrupt "what?" comes from Celia and a puzzled "Eiko?" from Ben. Leo's mouth twists into a smirk.

Celia is the first to recover. "Eiko," she says, speaking slowly and clearly. "Eiko Carr, are you a youngtwin?"

"I am," Eiko answers, not looking up.

"But your mother–"

"Not my mother. My sister, Nariko." She draws a deep breath and turns to Celia. "She died twenty-four years ago, just short of five years old."

Mumbles of surprise and sympathy come from around the table. Celia puts a hand on Eiko's shoulder.

"It was an accident. Nothing genetic," Eiko continues. "Obviously. Or they wouldn't have…"

"Eiko, sweetie," Celia says, softly. "Why didn't you tell us?"

"I didn't know. Not until my reading. Apparently, my parents couldn't quite figure out how to tell me. So they didn't. And after the reading, I… I didn't feel like talking about it."

"Well, what do you know?" Leo's tone is unpleasant, mocking.

"Shut up, you twat!" Celia hurls at him. She turns back to Eiko and continues gently: "Have you spoken to your parents about it? They seem so…"

"No, I haven't. It's too… complicated."

"I can't believe they told you a thing like that at your reading," Celia says. "That's just *so* inconsiderate. Idiots."

"It wasn't like that," Eiko says. "My sister's DNA was in the system already. The councilor assumed I knew. He apologized profusely when he realized that I didn't. He was the one who told me about my sister's accident, to reassure me."

"Same thing here," Father Elias says. Celia and Eiko turn very quickly toward him, as if they had forgotten about him. "We have access to your DNA database and the perfect match simply popped up," he adds. "Unfortunately, this piece of information ties our hands." He pauses, looking around the table. Seeing that he has everyone's attention, he continues. "There is another piece of information that is highly relevant to your quest here. I told Ben himself earlier in the day, but I feel that you should all know—to help you understand." He pauses again. "Ben is pure."

"He is *what?*" Leo says.

"Pure. He is one of us. Unedited, as you call it." He pauses briefly. "Unfortunately, Ben is not…" The rest of Father Elias' words are drowned out in the barrage of questions and exclamations flying across the table.

"What the hell…" comes from Leo. He leans back in his chair, staring first at Ben, then around the table, ending on Father Elias. His gaze is not returned by any of them.

"Ben, you red-haired rascal!" Celia exclaims, smiling at him. She turns to Eiko and continues. "Eiko, did you know about this?" Eiko starts to answer, but it all gets tangled up in her head: the then and now, the guesses and the knowledge, what not to say and what not to show. Celia tries to help her along with additional questions and general reassurances.

"How is this even possible?" Raphael slides forward again and leans eagerly over the table, toward Ben. "Did your parents cross over? Did you know? The accident… I thought… And what about your reading?" Raphael's questions pour forth. They are equal parts concern for Ben and simple curiosity. Ben is happy to engage his friend's interest. It has been too long since they last talked in any depth.

After some time, Leo speaks out, addressing the table as a whole. "You *do* all realize that Father Elias has left the room—don't you?" The four others turn to look at the empty chair. They acknowledge the truth of the statement, but do not seem much affected. "Without saying goodbye," Leo adds and snorts. The conversations restart quickly and flow eagerly, constellations shifting along the way. They stay and talk until they are called for dinner.

*   *   *

The moon is out, thin and crisp like a curved sword. It gives just enough light for Celia to see the person standing on the lawn, his back turned, and to guess who it is.

"Amazing, isn't it?" she says. Ben turns around with a jolt. "Sorry," she continues, "I didn't mean to startle you." She moves closer. In the dim light, the red of his hair is dark gray, making him look older, more serious.

"It's fine," he says. "I was just a million miles away." He can just about see her face. "I was trying to work out the constellations." He looks back up to the sky. "But you are right: It is amazing out here. So peaceful."

"Do you know any good ones?"

"Good ones?"

"Constellations." She looks up as well.

"Nah, just the usual ones: The big dipper," he points, "Orion and his belt," he points again, "that kind of thing."

"Another interesting day today, huh?" She says, after a while.

He looks at her, but only briefly. "Yep."

"Can I ask you something?" She says, still looking up.

"Sure."

"Eiko told me some things about your parents' accident—and the uncertainty surrounding it." She stops and glances over to check his face. He does not look back or visibly react. "Do you think it's connected to this? To your... status?"

"To them, and me, being unedited, you mean? I don't know. There are so many unanswered questions."

"And are you planning to answer them?"

"Wouldn't you?"

She does not respond directly. Instead she says, her voice lowered: "Eiko is afraid you'll cross over."

"I'm not al..." He stops. "It's a big decision."

"She's worried about you, you know? Really worried. I can tell. Even with all her own shit to deal with..." Celia sighs. "I still can't believe her parents didn't tell her earlier. I mean..." She sighs again. "Well, least it wasn't the usual selfish bullshit. The twinning, I mean. It must have been... grief or something..." She stops talking. He remains quiet. "Anyway, Eiko *really* cares about you," she continues. "That's all I wanted to say." Another pause. "God, what a crazy month, huh? Everything has been thrown right at us. You are adults now. Deal with this, deal with that..."

"Yes," he answers, adding nothing more. After a while, he turns to look at her, standing so very close to him. She is gazing at the sky again. He draws a deep breath and asks, his voice slightly shaky: "And you, how do you feel?"

"I'm fine." She shrugs. "I mean, yes, I've got a few extra edits, but I sort of knew about it beforehand. No major surprises."

"I mean about... How do you feel about..." He shakes his head and looks up again. "Never mind."

She keeps looking at the sky, as well, waiting for him to say more. He does not.

"She really cares a lot, you know?" She finally repeats.

"I know," he says, and again, more softly. "I know."

They are quiet for a while.

"Look," Celia says, suddenly, and points. "A shooting star."

*  *  *

"He's gone," Eiko says, her eyes full of panic. "I tried his room. It's empty."

"Could he have gone for a walk?" Raphael asks, spoon in the air. He is still working on his breakfast. "And packed up first?"

Celia is sitting next to him, closer than usual. She looks at Eiko and frowns. She is pretty sure she understands what has happened with Ben. But what she could say would not help anyone, least of all Eiko.

"Where is everyone?" Leo says with annoyance as he comes through the door. "I thought we said nine o'clock." He sees Eiko and finally seems to register what she has just said. "Benito, you fool," he says, mildly.

"We have to get hold of someone," Eiko says. "It's not safe for him in New Eden. We have to stop him."

"Look, if this is what he wants," Leo says, "then we have to respect it."

"He can't possibly want this," Eiko says, fiercely. "Not seriously. Father Elias filled him with all kinds of sentimental crap about community and family and ready-made purpose in life. He probably hinted that Ben's parents would be there, as well." She paces the room while talking. "No wonder he fell for it. He was even told that HQV was a thing of the past—*and* that its effects had been exaggerated." She shifts to a mocking tone. "Don't worry, Ben, there is no *real* danger." She throws up her hands. "How can they just say a thing like that? How can he possibly believe it?" No one answers. "There might be other stuff he needs, as well: cancer prevention edits, whatever. He has to get those somatics done soon. He can't just…" She stops, breathing hard.

"He can get them when he gets back, can't he?" Raphael asks, quietly. "Once he has gotten this… this adventure out of his system."

"Adventure?" Eiko's voice is shrill. "Seriously? Who's to say he will come back? That he *can* come back? That they'll let him? No, we need to do something—now!" She turns around and marches toward the door. Leo moves out of the way, revealing Father Elias in the doorway. Eiko goes straight up to him. "Do you know what you've done? What you made Ben do? You need to stop him."

"I think we should talk—before you all leave," he says to the room at large.

"No, we need to find Ben. Now." She continues to glare at his face. Then her expression starts to change. Her fury wilts, her eyes sadden. "*You* need to find him. I'm not allowed in, am I?" The full truth of her predicament hits her with another slight delay. As of yesterday, Ben knew this as well. She loses steam completely, then starts to cry. Celia jumps up and runs over to her.

"There are things you should know about the current state of affairs in chartered countries," Father Elias continues, disregarding Eiko's outburst.

"You *will* want to hear this." He looks at each of them, in turn. "I'll be in here." He turns around and walks into the blue room.

"This should be interesting," Leo says, following him.

Raphael picks up his cup and moves toward the door, as well. He stops briefly in front of Celia, who is still embracing Eiko.

"We'll be right there," Celia says.

"I can't," Eiko says, into Celia's shoulder. "I don't want to listen to that man, that evil…" she continues, angrily, then starts to cry again.

"You can do it. I'm right here with you." Celia straightens up, taking Eiko with her. She continues to mumble reassuring words as they move toward the blue room.

Father Elias wastes no time. "Ben left the transit station some time during the night. We don't know where he is."

"I thought you had a tracking system," Leo says. "Or was that just a bluff?"

"We think he may have slipped into New Eden. I want you to know that we did not encourage or facilitate this. He was not cleared to leave."

"Then how?" Celia asks. "Someone must have helped him." Out of nowhere, she thinks of the cook, Hetty. She keeps this to herself.

"Maybe he just walked off," Leo suggests. "The perimeter is not exactly tight out there."

"There's a wall, you know?" Celia says to him. "In that direction, you'd need to…" She mimics jumping over.

"Oh, right."

"Why wasn't he cleared?" Raphael asks. "I get that you are not too keen on the rest of us 'impure' rule-breakers. But given what you told us yesterday, Ben should have a free pass. Shouldn't he?"

"The problem wasn't on our side," Father Elias says, ignoring Raphael's sarcasm. "Your authorities didn't clear him to leave."

"He's been flagged as unedited in Med-data," Eiko explains, in a flat voice. "They'll want to quarantine him. Especially now that he's been in contact with you guys." She stares hard at Father Elias.

"There may be more to it." Father Elias speaks slowly. They have to wait for him to continue. "Ben is a rare commodity now."

"A commodity?" Celia says, scornfully.

Father Elias thinks for a while. "Do any of you know what is *really* happening at the reproductive services right now?" he finally says. He looks at Eiko first.

"I guess we don't," she says stiffly, after exchanging glances with the others. "I'm just a student helper this summer. And yes, my parents also work there, in research and the clinics. But they haven't told me anything. They don't always…" She purses her lips and looks away, stubbornly.

Father Elias turns his gaze to Leo. He waits.

"It's a big fucking mess right now," Leo says, unable to resist a smile. "My hack made it impossible for them to…"

"Your hack?" Raphael and Celia exclaim, in unison. "You hacked the Reserve as well?" Celia adds with amused disbelief.

Leo nods and lifts an eyebrow. "You smug little-" Raphael starts, lurching toward him. Leo reacts quickly and evades him. "Why would you screw with people's lives like that?" Raphael continues. "What's the matter with you?"

"The hack will eventually be fixed," Father Elias says. "But probably not for a while. The leadership will pretend it is endlessly complicated and terribly difficult—and secretly be grateful to a keen hacker for giving them cover."

"I did not-" Leo starts, furiously. Then he frowns and looks away.

"It doesn't matter whether you knew or not. The hack happens to be very convenient."

"You've lost me completely," Celia says. "Cover for what?"

"Yeah, what?" Raphael adds. "We have some idea of how the Reserve works. A serious hack can do a lot of damage."

"Yes, but in this case, it also masks the real problem."

"And what *is* the real problem?" Celia asks. "In your opinion?"

"You really don't know? Not any of you?" Father Elias seems genuinely surprised. "You don't know why the Reserve is—temporarily—closed for business?" Heads shake, slowly. "Or even *that* it is closed?" More headshaking. After a moment, Father Elias joins them. "That misinformation system passing for news in your world really works, doesn't it?" He sighs. "You hear everything, but know nothing. I'm not sure who should be more troubled by this, you or us."

"Look, we are just students," Raphael says, irritably. "We've told you we don't know. If you have something to tell us, just tell us."

"And don't forget Ben," Eiko says. "You are all forgetting Ben."

Father Elias clears his throat. "Fertility in chartered countries has fallen suddenly and catastrophically," he says. "No one knows why." He pauses. "You might not have been paying attention to this, at your age. And it has been kept very quiet."

The four look at each other, surprised and worried, but also skeptical. Except for Leo. Something else creeps into his face: Satisfaction.

"We don't need a high fertility rate," Eiko tries. "We have procedures…"

"It's practically down to zero," Father Elias says, matter of fact.

"But you said the Reserve is closed," Raphael says. "So which is it?"

"Temporary closure because of 'IT problems' buys them time. They are using it to conceal the more serious problem, the fertility block."

"How would *you* know about all this?" Eiko says, her anger at the priest resurfacing. "And what does it have to do with Ben?"

"The pure don't have a fertility problem. Only the impure." Father Elias pauses. "Sorry," he adds with a minimal smile, "I mean the edited. Only the edited have this problem. All chartered countries are affected by the fertility block, even if they have yet to admit it. You are all affected, all sterile." He makes a sweeping gesture, his hand, for once, bared. "Not Ben, though," he continues. His hand is re-cloaked. "He is healthy. Of course, he is just one person, one male…" He waits for this to register. It does. Celia and Eiko both look up, with distaste. "Perhaps now you understand why we expected you to be coming to us with a proposal, with something that could initiate informal negotiations."

"Us?"

"Well, obviously, your government needs our help—and needs it very badly." Father Elias speaks slowly, taking his time. "A visit from the children of people in power could be seen as a signal. Plus, anything that you might have proposed would be deniable in future negotiations, if needed." His words provoke only blank stares. "It seems we may have read this all wrong. Maybe the five of you turning up here, making contact with us at this particular time, was just a coincidence." He sounds quite unconvinced. But then his expression becomes less certain. "Admittedly, we never did understand how your friend Ben fit into the picture, so…"

"But the recent political changes?" Celia asks, finally paying full attention. "The restart of bilateral talks, the possibility of visits, these transit stations," she gestures at the space around them, "that is *not* a coincidence, I presume."

"No, it is not. The fertility problem became clear to a number of people in your world—and to those of us with good sources—some time ago. Thanks to Leo's hack," here he looks at Leo, who glares back, angrily, "your populace can probably be kept in the dark for a while longer." He pauses. "But eventually, your government will realize that there is only one way forward." Another pause. "And eventually, they will embrace it."

"So you want us to transmit this threat, or whatever it is, to our respective parents? Is that it?" Raphael asks.

"The people who need to know, they know already. They also know about us. Sooner or later they will—well, we shall see." He smiles. "Interesting times." He gets up from his chair and nods briefly to each of them. "But maybe it is best that you return to your homes now. There is nothing more to be accomplished here."

"What about Ben?" Eiko says.

"Ben will find his own way. And I think we should let him." Father Elias makes a shallow, formal bow and exits the room soundlessly.

*   *   *

"Where is Leo?" Raphael says, irritably. "I thought he wanted to take off as quickly as possible." He, Celia and Eiko are standing in front of the pod. Celia, like Raphael, looks restless. Eiko looks sad and resigned, the panic long gone.

"Is it true, do you think?" Celia asks them both. "What he said about a fertility block? I can't think of anyone who is actually pregnant, but…"

"It's easy to find out," Raphael says. "Once we get back, we can conduct a real-life survey." He turns to Eiko. "Eiko, if he's right, lots of people at the Reserve must know that something is off. Are you sure you haven't heard anything?"

"I really haven't," Eiko says. She furrows her brow. "Well, nothing specific, anyway. My parents have been complaining about a major new project soaking up all the resources. They haven't told me what the new project is about, though…"

"I bet once we start looking, we'll find the stories on the net already," Raphael says. "Something like: 'Catastrophic decline in fertility threatens mankind. Government cover-up suspected.'"

"There are stories about every possible thing, all the time. How are you supposed to know?" Eiko asks, irritably. Raphael shrugs. Eiko regrets the misplaced reaction immediately. "I'll try to find out what I can," she adds.

"And I'd better ask my mother," Raphael says, with a sigh. "Of course, there'll be Leo's hack to complicate matters. Assuming he really did that."

"I think he did," Celia says. "He was furious when he realized how the authorities could use it to their advantage. That was real."

"I don't get why Ben made friends with him," Eiko says.

"Leo seems very protective of Ben," Celia says. "He's probably been a good friend these past couple of years."

"I wish we had done more to stop him," Eiko says. "Ben, I mean. We could have talked some sense into him. If we had explained…" She sighs. "Did either of you talk to him last night? Before he left?" She turns to Raphael. "Rafi? He always listens to you."

"No, sorry," Raphael says. "Too much excitement yesterday. And Leo's bottle last night—strong stuff. I went straight to bed after that. Slept like a log."

Celia doesn't add anything.

They wait, all three facing the front door of the building.

"I can't help wondering why Father Elias told us all that stuff—assuming that it *is* true," Celia muses. "He didn't have to."

"He did it to gloat," Eiko says, with bitterness.

"Maybe when he realized we were not sent here as emissaries," Raphael says, "he decided to use us to boost the reverse information flow."

"Makes sense," Celia says. "Spread the word, light a few fires."

"If he's right," Eiko says, "the panic will spread all by itself—as soon as it becomes public knowledge."

"If he's right," Celia adds, "New Eden is sitting on a goldmine. I'm pretty sure they'll want to take advantage of that."

"Scary thought," Raphael says, shaking his head.

Just then, Leo comes striding toward them, full of energy. He unlocks the pod. "Sorry for the delay," he says, smiling, as he gets close. "There was something I had to take care of."

Eiko climbs in the back. Celia follows. Raphael smirks at Leo when he sees this. But Leo seems not to notice or not to care.

No one is there to wave goodbye.

·

# Part II

# 5

## Home

Ben pushes the wheelbarrow along the well-worn track. Being upright and walking comfortably is a relief after the strenuous morning in the field. He settles into a good pace. The wheelbarrow is not heavily laden, just a bucket of the new potatoes. He feels the warmth of the sun on his back and the cool of the wind on his bared stomach. He smiles.

Ahead of him, Ben sees Harry from the neighboring farm, along with a small herd of brown-and-white cows. Harry is twelve, but small for his age. He has opened the gate to the field on the right and is directing the large animals onto the fresh grass. A couple of them display un-cow-like friskiness, running into the field with ungainly abandon. One maverick seems to find the path ahead of her more interesting that the intended field, so Harry has to run around her to set her right. He does this with a slender stick and a confidence that Ben finds astounding. After completing the transfer, Harry notices Ben and raises his hand in greeting. Ben reciprocates. The spontaneity of Harry's gesture and his own easy response makes Ben unaccountably happy.

The transfer reminds Ben of the summer when he was herding bots around to clean offices. There was a maverick in that pack as well. This bot responded to openings between rooms not by going through, as intended, and starting on another well-calibrated up-and-down floor cleaning routine, but by veering off to one side and speeding up, as if enjoying a burst of freedom. He knew that he should ask for a replacement bot, or at least a reprogramming, but he found the necessary coaxing a fun distraction in an otherwise boring job. That summer—the early-morning job that felt strangely important to him, the day of the reading, Aunt Vera's letter and the bizarre trip to the

© Springer Nature Switzerland AG 2020
P. Rørth, *The Unedited*, Science and Fiction, https://doi.org/10.1007/978-3-030-34624-9_5

transit center with his friends—was that really only two years ago? His other life—before. It seems so remote now.

Harry has run off some time ago. Ben realizes that he needs to get a move on. He should be helping Tilly in the kitchen by now. They have a lot to do today. He should also get out of the direct sun. From the field several pairs of large, docile eyes observe him putting his shirt back on. He enjoys working in the kitchen, he tells them. He knows his cousins, Anne and Peter, as well as Peter's friends, find it ridiculous. But that doesn't matter. They probably find him ridiculous, but that doesn't matter either. Tilly matters, though. He waves goodbye to the cows and continues along the path with increased speed. The potatoes jump about in the bucket, but do not escape. He can almost taste them now. They are so amazingly good. He must be hungry.

"Ben, is that you? Did you remember to bring extra potatoes for tomorrow?" Tilly is yelling through the open kitchen door. "Our day of rest," she adds as she appears in the doorway. She gives him a warm and sunny smile.

"I think I've got enough," he moves the wheelbarrow into position in front of the door. They inspect the bucket together.

"Yes. That looks fine." She picks up the bucket and, with a practiced move, swings it into the large rough sink inside the kitchen. She winces from a sudden pain, but makes no sound. Ben has already turned away and does not see this. He rolls the wheelbarrow to the shed and returns immediately.

"Where do you want me to start?" He says with a big grin.

"How did I ever manage without you?" Tilly raises a hand to ruffle his curls. Tilly's hair is a faded auburn with some gray in it. Ben guesses it had the same color as his when she was young. He does not jerk his head away as a boy of Harry's age might. Instead, he leans into the caress for a brief moment. Then he reaches for the large pot behind the door and gets to work.

*    *    *

"Ben, I need to tell you something." Tilly sounds serious. "Let's sit down." She indicates the wooden bench by the long worktable. It is after the midday meal and they are back in the kitchen, getting ready to set up bread for the evening and for Sunday. Milly is there as well, silent and watchful. Ben and Tilly sit down next to each other at one end of the bench, both in their aprons. Milly sits at the other end and places her favorite doll, soft and worn, next to her. They both have aprons on, as well.

"I got some news this morning," Tilly continues, wiping her hands on her apron, "from Father Len who got it from Father Marius."

"Who is Father Marius?"

"He's an old friend of the family, who has been helping us look around. He knew your mother, as well."

"And my father?"

"He knew Jacob less well. But he knows where his family is at. He has even been to see them—or maybe he sent someone—I'm not sure." She pauses.

"What did they say?"

"Well, your father's family has not changed, I'm afraid. They still say that Jacob is no son of theirs, not anymore."

"So I guess I'll never meet them." Ben feels disappointed, as he knows he should, but also a bit relieved. "I don't mind. I'm happy here."

Tilly smiles. A moment later, she continues. "But that wasn't what I wanted to tell you."

"No?"

"No. It seems that Father Marius has heard back from the last of his enquiries." She straightens up and turns to face Ben directly. "I'm afraid there is no trace of Jacob and Eleanor anywhere in New Eden, under the new names or the old ones. There's been no sign of them since they left."

Tilly turns sideways again and places both her arms on the table. She sighs.

Ben looks at his clasped hands for while, then back at Tilly. "Do you think he has looked—or asked—everywhere?"

"I do. Father Marius has very good connections. He works for the archbishop in Petersburg and he's an authorized contributor to KnowledgeNet. According to Father Len, this means he is able to contact all the parishes directly." She nods a few times. "I know him, Ben. He's a very decent man."

"I see…" Ben sighs. "So this is probably it."

"I'm afraid so. Father Marius was our best chance and he has done what he can." She looks at Ben with sadness in her eyes. "I think the accident was real and the sea simply took them. I *am* sorry, Ben."

"I suppose I sort of knew by now," Ben says after a while. "It's been two years. Well, no, four years. They would have made contact if…"

"Yes." She waits. Ben does not add anything. "Maybe I could ask Marius whether…" She catches herself. "No, I think he tried everything."

"Did you tell him why you were looking for your sister now, after all these years?"

"Yes, of course. I told him about you coming home—and why you still had questions about your parents." Ben looks uneasy. Tilly notices. "Don't worry," she adds quickly. "He's high up, but a true friend. I trust him. Everyone around here knows about you, anyway."

"Yes, I guess they do." Ben has a sudden recollection of arriving at the farm. The first few minutes had brought gut-wrenching panic and an awful certainty that he had made the mistake of a lifetime. Then Tilly came out of the house and somehow just *saw* him. She gave him a hug without asking first and immediately started organizing his new life. She set everyone straight about who he was and where he belonged. Neither Harold, Tilly's older brother and now the head of the family, nor his wife Daisy seemed inclined to dispute Tilly's authority with regards to orphans. Tilly also told anyone interested that he was nineteen. She cut short his protests, so he understood this was for the best, somehow.

In those first weeks, the children of the village would stop and stare when he walked by. They were a little afraid of him, Tilly said. Some of the parents started whispering that the newcomer was impure, possibly dangerous. Father Len put a stop to that. He has been great. The children still think he talks funny, but they are no longer afraid of him. Now they ask him wide-eyed questions about life over there. Is it true that everyone is two meters tall and lives forever? Is it true that babies are cut into small pieces, grown in plastic bubbles and have no mommies? Do they know that they will go to Hell when they die? They don't seem bothered by the contradictions. They also don't listen to his answers—or they don't believe him. The hardest part is when they ask about the infamous clones. Has he met any? Are they terribly scary? When they are not asking questions, the children like to explain things to him. Everyone has been much amused by his ineptitude when it comes to outdoor work. He has gotten better, though, just as he has gotten better at finding appropriate answers to their many questions.

"Tell me about my mother again," Ben says, his mind back in the kitchen. He makes himself sound extra eager, like a small boy asking for a bedtime story. Milly appears to notice the switch immediately. Grasping her doll, she slides down the bench to join Tilly and Ben.

"What do you want to hear?" Tilly has shifted to a jolly tone as well.

"Tell me about how she first met my father. At the harvest festival."

"I'll tell you the story, but we have to work while I do."

The three of them line up at the long table, the bench pushed halfway under. Tilly and Ben each place a large bowl in front of them. They position the jars of flour and salt and a jug of water between them. Milly gets a smaller bowl, placed on the long bench. Ben goes to pick up the sourdough from cool storage.

"Well, first of all, it was in the springtime," Tilly says.

"Oh, yes, the planting festival." Ben places a large dollop of sourdough in each of the big bowls and a small one in Milly's. Tilly measures out flour and water and pinches of salt while she continues talking.

"Ella was a vision of spring, her hair speckled with flowers. It was the exact same color as yours, but long and very frizzy. It was perfect for trapping daisies, buttercups, cowslips and any number of other flowers. Petite and graceful, she looked like an angel."

Before digging into their bowls, they line up at the sink where Tilly first uses, then passes on, the large bar of soap. Ben holds it for Milly to rub at. When they have finished this part of the ritual, they return to the table and the bench. They start mixing dough with their bare, and soon very sticky, hands. Tilly continues to describe her sister, dancing and laughing and enchanting everyone at that long-ago spring festival.

"Tell me about my father," Ben says, when the flow of words becomes a trickle.

"Jacob's family lived far away from here. He was visiting a cousin, I think, or someone he knew from the guards. It was rare for anyone to travel—it still is, you know—so any stranger attracted attention. He was darker, and leaner, than most of the men from around here. He looked exciting—even a bit dangerous." Tilly makes big eyes at Milly, who giggles soundlessly. Tilly goes on with the story. It is as close to the truth as it needs to be and as good a story as to be worth retelling. For Tilly's select audience of two, it has already become a favorite. In the meantime, the sticky messes are transferred from the bowls to the table and gradually become perfect spheres of elastic goodness—two large and one small.

*   *   *

The next morning carries the promise of another fine day. The family is assembled in the courtyard, but no one speaks. The moment the bells of Hatton village church commence their summons, they start moving. Ben always walks at the rear, next to Tilly. Harold and Daisy lead the way, followed by Peter and Anne. Although close to him in age, Ben's cousins almost never speak to him. This puzzled him in the beginning, even hurt him, but now he finds it a relief. Peter is quiet and obedient at home, wary of his father's hard hand and harsher words, but with friends he tends to be boisterous and unpleasant. Anne has a prominent scar on one cheek and is painfully shy in all company, not just Ben's. Their mother rarely speaks to anyone, as far as Ben can tell, except to give her daughter orders in a clipped voice. She spends much of her time praying and the rest of it cleaning the house, over and over. Tilly seems to have convinced both her and Anne to stay out of the kitchen. She runs the family kitchen as efficiently as she runs the local school. The school sits every afternoon during the week. Tilly corrects homework at night. Ben sometimes worries about her not sleeping enough. She claims not to need it.

"Mathilda," Harold says without turning around, "do we have a decent Sunday cake ready for after the service? I was planning to invite Father Len to join us." According to Tilly, Harold regards nicknames as an indulgence best never to get the taste for. On the rare occasion that Harold addresses Ben he calls him Benjamin. Ben does not correct him. Harold never mentions his sister Eleanor.

"Yes—fresh out of the oven," Tilly answers. She and Ben exchange amused glances. It tickles Tilly that Harold feels compelled to invite the local priest over regularly, even though he does not seem to like him. Daisy's silence becomes even more tense and tight-lipped when Father Len comes around. She obviously disapproves of him. Tilly likes him, as does Ben. He is well read and interesting to talk to. Initially, Ben had been worried that the priest would chastise him for his lack of faith. But that seems not to be Father Len's way. The first time they spoke in private, he asked Ben to continue taking part in church and community activities and to be respectful of their faith. Nothing more was required. If Ben would be open to occasional conversations about spiritual matters—or any other matters—Father Len would be delighted. Saying this, he had smiled—a relaxed and generous smile. Returning to a more formal tone, he had added that Ben must *not*, at any time, transmit foreign or seditious ideas to children or other susceptible souls. He somehow made Ben understand that he, Father Len, was obliged to report any signs of such subversive activity—but also that he would very much prefer not to do this. One-on-one they discuss anything and everything.

Milly has fallen behind. Tilly notices and gestures to Ben. He goes back and asks her if she wants to be carried. She is slight for her six years, so it is easily done. She nods a yes, but still looks unhappy. He guesses that the other children at Sunday school tease her. Being both an orphan and unable or unwilling to speak, she is an easy target. With Milly securely on his back, Ben walks quickly to catch up. He enjoys the weekly service much more than he had expected to and does not want to miss the start. The church itself is a quirky old stone building, gray and a bit dark, but with stained-glass windows that let in plenty of color and light. And, most importantly, the church has an organ. It is not very old or very fancy, but it is fully functional and the organist is obviously talented. Ben has spent enough time at Eiko's house to be able to tell. The organ and the organist attract good singers from an extended area to the choir. Father Len is well aware what a treasure this is for the church. He makes sure music is a significant part of every service. A few of the choir members even sing solo pieces. Ben figures he may now know what heaven sounds like.

The only awkward part of the service is when it is time for communion. They go forward pew by pew, row by row, and Tilly always takes part. Ben stays behind, uncomfortably exposed, and tries not to look around. When Tilly returns she sends him a reassuring smile—every time.

\*  \*  \*

"Thanks for passing on the information from Father Marius," Ben says to Father Len as they are leaving the church. The others have gone ahead. Father Len always stays after the service to exchange a few words with any parishioners who may need his ear. Today's service was well attended, but luckily no one lingered for too long afterwards. Ben helped with the tidying up.

"I'm sorry it wasn't more positive," Father Len says. He is at least a head taller than Ben, but even skinnier. The loose frock hides the details. He stoops slightly and has a scrawny beard. His age is impossible to tell.

"I wasn't expecting anything else at this point," Ben says and looks at his feet. "Hoping, maybe, but… I appreciate everything you've done to help."

"Thank Father Marius," Father Len says. He frowns briefly. "I'll thank him for you, if you want. He's not easy to get hold of."

"Please do."

They walk on down the lane that takes them in a southeasterly direction, enjoying the sunshine and the bright, fresh green of the trees. Father Len insisted they take the usual detour. He needs the exercise, he says. A couple of robust workhorses are appreciating their day off in the field. Ben remembers the horse he saw earlier. It was of a different ilk.

"Who was the mounted uniform outside church today?" he asks. "Before the service. Peter and his friends seemed to know him."

"He looked like a colonel, from the guards," Father Len answers. "The boys probably trained with him."

"Oh," Ben says, looking away. The information reminds him of his recent encounter with the three friends. Ben had come out for a walk after an afternoon in the kitchen. Peter, Mark and Gustav were sitting on a low wall and talking in an animated way. It was about girls, Ben could tell. The path that he was on went right past them. It would have been awkward to turn around.

"So, are you coming to the dance this year?" Gustav asked Ben as he passed them. He has always been the friendliest of the three and almost certainly meant no harm. Ben stopped and turned to answer, but hesitated. He had not known about the village dance the year before—somehow everyone had forgotten to tell him. The first he heard of it was when Tilly, working her way through a stack of exercise books, asked him why he was at home and not at the dance. She had not yet realized how unhelpful Peter was being to his new cousin. This year Ben was better informed, but he felt ambivalent about the upcoming event.

"He's waiting for a nice girl to ask him," Peter said before Ben managed to find a reply. "Someone he can swap recipes with."

"Good one," Mark said, smirking.

"I'm going to wear my uniform," Gustav said. "The poster encouraged it this year."

"Me too," Peter said.

"Alright. Me too, if you insist," Mark said, nonchalantly. "I'll dazzle those poor girls with my honor strips."

The three of them had slipped off the wall, one by one, and were now standing up straight, as if talking about uniforms made slouching inappropriate. They turned to Ben, who was feeling more and more uncomfortable.

"Do you have guard service—over there?" Gustav asked.

"Not mandatory service, no." Ben considered whether to explain about citizens' duties and the dedicated education for defense service jobs. He decided not to. He had a feeling it would not be well received. Mark, Peter and Gustav soon got restless and started rounds of mock combat: Bursts of rapid forward movements, boxing morphing into wrestling, and shouts of perceived wins. Mark tried to engage Ben, but Ben simply stepped backwards. They did not bother him further. It was not necessary. He was quite fit now from the daily manual labor, but still significantly lighter than any of them—still a boy. The beard he had grown since he came had not made any difference. It was sparse and orange-red in color. More convenient than shaving, but that was all. Some things never change, Ben thought as he moved on. He was sure they were having a good laugh behind his back. He did not turn around to check.

"There is nothing wrong with not having been a guard," Father Len says, back in the present. "It's mostly just a waste of time." A hint of worry crosses his face. "But don't quote me on that. It's not a popular sentiment these days."

"I won't tell a soul. Especially not Harold," Ben says and they share a look. "Disrespecting the guards might mean the end of Tilly's Sunday cakes."

"Now *that* would be tragic."

They both smile and then walk on. After a while, Ben speaks again. "What was he doing here, do you think? The colonel. I've also heard they are encouraging uniforms at the village dance."

"Don't worry. Most people will be dressed as usual."

"But is it significant? Has something happened in the outside world?" Ben pauses briefly. "Sometimes I feel like we're living in a bubble. There could be a war going on and we wouldn't know about it."

"We would, though. It would be on KnowledgeNet if something was actually happening." He pauses. "At this point, all we have is rumors."

"And?"

"There's some tension in the air: Increased guard presence at the wall, older ex-guards being called in for refresher training. For us, more strategy meetings and increased pressure to submit diligent reports." He looks at Ben. "Reports of possible subversives." He sighs. "This kind of thing has happened before. It means people at the top are worried about something."

"Do you know what they are worried about? Or are you not allowed to tell me?"

"All I know is that there is trouble on the other side. Unrest, they call it. I guess they are worried it will spill over the wall."

Ben almost asks if Father Len has ever seen the wall. Nothing can spill over *that*, he thinks. "What kind of unrest?" he asks, instead. It comes out sounding skeptical. He can't help it. Unrest in a chartered country just doesn't seem very likely.

"I don't know. I'm just a village priest."

"But you have KnowledgeNet access."

"So do you. You just have to go to the school and ask to use the terminal. Did I not tell you about that? Or Tilly?" Ben shrugs, not sure if they did. "We should have," Father Len continues. "KnowledgeNet has lots of information about all kinds of topics, including current events and recent history. Some of it is really interesting, even if the presentation is a bit dry." He chuckles. "I guess it's nothing like your net."

"Which is anything but dry: fast-paced output of endless trivia. Anything goes, as long as it entertains. You wouldn't believe how much junk there is."

"You've told me about it. But, I admit, it *is* a bit hard for me to imagine."

"It's not worth imagining."

Father Len looks pensive. "You really don't miss it?" he finally asks. "The net. The entertainment." He holds the last word apart, as if unsure whether it is a real thing.

"Not really. When you watch it, it's like you disappear—or time disappears—I'm not sure which." Ben frowns. "It's a different kind of disappearing than when you read a book or listen to music." He frowns again and shakes it off. "By the way, I really loved the last piece today."

"Bach," Father Len says. "Good for letting the mind roam free."

They walk side by side for a while, without talking.

"But I *am* getting curious about what's going on," Ben says. "Out there, I mean." He adds a sweeping gesture.

"I get that."

They pass a small copse of oak and birch and the farm is now visible, close enough to see an open door and someone standing in it.

"That's Aunt Tilly, I think," Ben says, "keeping an eye out for us."

"Your aunt is a very special woman."

"She is." Ben smiles.

"The school—the children—they are lucky to have her."

"They are."

"Have you ever considered going in that direction?" Father Len asks, tentatively. "I've seen you with little Milly and it seems—well—that you have a talent there. I could put in a good word with the academy." He looks over at Ben. "That's assuming you stay, of course."

"I don't..." Ben returns Father Len's gaze. His expression is serious. "You know? I've never thought that far." He pauses. "Do you think the parents would accept me?"

"Everyone really respects Tilly."

"She's not an outsider."

"Neither are you. Not anymore." Father Len sounds like he believes it. "It's just a thought," he adds.

"Thanks."

Tilly sees them and waves.

"Now tell me, what treats might we expect to sweeten our conversation with today?" Father Len's voice goes up a notch. "When approaching Tilly's kitchen, I find indulging in anticipatory speculation a most satisfying pleasure." The smile lights up his whole face.

\* \* \*

"She looks comfortable enough," Tilly says as they look at the sleeping figure of Milly on the floor. She is lying on some stray pillows in the corner of the kitchen. "I think she prefers to be wherever we are."

"She does," Ben says and returns to the sink area with his dishtowel. "I wonder if she will ever talk."

"I think she will, when she's ready," Tilly says, sitting down. "Her hearing is fine. If she can't speak, she will write notes instead. She can read already."

"She can?" Ben turns around and tilts his head, quizzically.

"I used to follow along with a finger when I read to her. Now she does it." Satisfaction fills Tilly's face. "And she looks through the books by herself—even those without pictures."

Ben puts the last dishes away and joins Tilly at the long table. "Impressive," he says and takes the cup of tea she has poured for him.

"I do worry how it will be when she starts school," Tilly says. "A classroom is a very complicated place."

"I think the other children tease her a lot."

"They do," Tilly says, with regret.

"Do you know anything about her parents?" he asks, after a bit.

Tilly pulls a strained smile and shakes her head. "It's better this way," she says. He does not try to push her. He knows it won't work.

They sit in silence for a while longer.

"Father Len," Ben says, "did he know my mother?"

"No. He came down from Petersburg ten-twelve years ago."

"Petersburg?"

"Yes. He came here straight out of training, I think. I always wondered how he ended up with us. But I am happy that he did." She smiles. "Why do you ask?"

"Nothing, really, just… I'd like to know more about my parents: what they were like when they were young, why they left—and what actually happened to them." Tilly looks at him steadily, but does not answer immediately. "Now I know for sure that they are gone and that I'll never be able to ask them directly," he continues. "Please, Aunt Tilly, tell me what happened back then. I should know."

"I don't know all that much, really. I don't know how they crossed over or who helped them. I never heard from Ella after they left." She can't quite keep the disappointment from her voice. She looks at her hands, wringing them slowly.

Ben waits for a while before responding. "I'm sure she would have been in touch, if she could." He speaks softly. "Maybe it wasn't safe." He waits for another while. "You must have known her really well," he continues, carefully. "You were close, weren't you?"

Tilly nods. "We were. Just a year apart."

"Tell me about her."

She looks up at him. "What do you want to know?"

"What she was like back then. But the real story, not fairy-tale, like for Milly."

"Alright." She pauses. "But I need something in return."

"Anything. A month of doing all the washing-up, two months of correcting homework," he makes a face "even math."

"Not that." She smiles, gently. "I want you to tell me more about their life later on… how they managed in a foreign world, what they were like as parents, as middle-aged lovebirds." She smiles. "I missed all those years."

"The Bella and Jack Hatton years."

"Yes. You know, I think I remember Jacob, or Jack—I like the name Jack, by the way—calling my sister Bella. And she was… very beautiful: a pretty face, but also that gorgeous hair, a certain sparkle, everything about her. That's not fairytale. She must still have been beautiful, later on…"

"I suppose. Her hair was almost always—tucked away, somehow. I think it might have been different when I was little. Loose. Maybe…" He frowns slightly, trying to remember.

"The long red hair went well with her wild streak. She often said that she hated it and that she would cut it all off some day. The idea drove our mother to distraction, as intended. I don't believe she ever meant it."

"She was wild?"

"Oh, yes, but with purpose, you know. She was also smart and very stubborn. Headstrong, I guess you'd call it. Our father did *not* approve." She shakes her head, slowly. "Harold is a lot like him, I'm afraid."

"Oh."

"But our mother wasn't at all like Daisy. She was strong—and had her own opinions. It's from her that Ella and I both got to love books." She smiles. "She took us to the library every week, even though my father didn't… She protected us. So it wasn't that bad. It was stormy sometimes, sure, but… We had a good childhood here." Tilly tells a meandering story about the forest and a stream and the two of them keeping secrets. It winds down and she switches up the tempo. "When we got older, it got more difficult. Ella would challenge our parents, especially our father, all the time. She would stay out late, talk back to them, hang around with boys, openly—the usual stuff for our age, I suppose. When our father tried to discipline her, she'd shake that fiery hair of hers and storm out, head held high." She sighs. "I was just one year younger but I had none of her proud defiance—or her self-confidence. I admired her, very much. To be honest, I was probably a little bit afraid of her."

"It's hard for me to imagine her wild and rebellious. But being afraid of her—that, I totally get." Ben grins. "She was always very strict with me—much tougher than the other Moms—about homework, about who I hung out with, making sure I helped out at home, everything. Mostly I just did what she said. Standing up to her seemed impossible, somehow. If I had only known…"

"But she did do brilliantly in school. She worked hard at it and always got top grades. It wasn't all about being a rebel. She was ambitious and she had goals. She wanted to become a doctor, you see. For some reason, she wanted that very badly. Our father wouldn't let her. He thought it was all Margot's bad influence."

"Margot?"

"She… I'll tell you about Margot some other time." Tilly frowns. "But that was what triggered it, I think. His refusal to support her ambition made her even more determined to break free, no matter the consequence. That and the rather stupid priest we had at the time. He equated faith with blind obedience. Ella loathed him. She started talking about leaving, about crossing over. Well, not talking, whispering. Crazy dreams, I thought. This was well before Jacob came along."

"So she was the one who made the decision to leave, you think?" Ben isn't really surprised. Bella giving orders was how it usually worked at their house. Even though he never protested, it used to annoy him, partly on his father's behalf. At some point, he realized that his father was perfectly content in his role. Then he felt betrayed and a bit ashamed. Now he can't help but feel proud of his willful mother, taking on the world.

"Absolutely," Tilly answers. "Ella was always the one in charge. But you know my story about how your parents met at the village dance? That *was* the way it happened, it really was. Your father was a great dancer and he did look awfully handsome and ever so slightly dangerous."

"I'm afraid I didn't inherit much of that from him—the looks or the dancing skills." Ben tries to imagine his father as a youthful charmer. It makes him smile. "Too bad."

"But you are so much like him." Tilly reaches out to touch his cheek, very briefly. "You are kind and generous. You are a good person, Ben."

"Thank you," he says, suddenly overwhelmed.

"You've become very precious to me. You know that?" She says, softly.

Ben nods but finds no words.

"She had him wrapped around her little finger in no time," Tilly continues, her voice firm again. "She could do that to any man. He didn't stand a chance. A headstrong beauty is obeyed. That's just the way it is."

Ben imagines his mother younger and stronger. He frees her hair, removes the lines of disappointment and bitterness. He almost succeeds.

"So, did Ella become a doctor, in her new life?" Tilly asks. "You told me she worked in a hospital."

"She didn't become a doctor," Ben says. He feels vaguely guilty about this but is not sure why. "She was a nurse—or a nurse's aide. I don't think she ever got the formal qualifications. But she was hugely popular at Sunset. They called her 'our angel', 'Bella, our angel', both the patients and the doctors. I heard them." He remembers the shock of it, visiting Sunset one weekend and seeing her as this completely different person: kind, gentle, but also amazingly competent.

"Sunset? Is that a hospital?"

"Sort of. A hospice."

"A place where people die?"

"Yes. She told me a lot of doctors and nurses found it difficult to work there. Many transferred out as soon as they could. She stayed. Along with the washing, feeding and medicating, she'd listen to their stories. She'd make them feel better, somehow. Dad said she had a special talent for making people remember more and fear less. The patients all asked for her. Even those who only checked in a few days before the procedure, they all wanted to talk to Bella."

"Procedure?"

"It's how most people die—when they choose to go." With a slight delay, Tilly's eyes go wide with alarm. Ben notices. "It's different there," he says, carefully. Tilly just nods. Ben scrambles to find a new direction. "Sometimes she'd come home at night and tell me about someone she had just talked to. She'd know their whole life story. She always spoke so fondly of them." He remembers feeling jealous, stupidly jealous, of these very old people. "She had all kinds of patients. One of them…" He stops and checks Tilly's face.

"One of them—tell me." She seems calm again.

"One of them was quite famous. She was one of the original…" He shakes his head. "No, never mind… Another one, Giovanni was his name, had made dolls all his life—amazing dolls, even though they were inanimate. You know, not robots. We went to the shop once." Ben goes on to tell the story of the doll-maker and the love of his life. The two had only been together for a short time and quite some years earlier. One of Giovanni's creations, a doll that keeps breaking, is a central character of the story. It surprises Ben that he remembers all these little details. "So we went there and got the doll for him," he finishes.

"That's sweet."

"The shop was incredible—all these life-like dolls everywhere. It was a bit creepy, too. That's probably why I remember the story." He frowns. "I've never seen Mom so emotional. I think she cried. In the shop. I never dared ask her why."

"You should have."

"Maybe, but…"

"I understand." Tilly pauses briefly. "And Jacob, no Jack—what did Jack do? Was he a doctor?"

"No. They never went to Uni, either of them. I don't-" He stops, realizing there is a trace of an old bitterness in his voice. He knows it shouldn't be there. "No, he was a cleaner."

"A cleaner? Really? But he seemed so…"

"I tried it, one summer. Just before I came here. You don't do the actual work yourself. You herd this clutch of robots around and they do the cleaning. It's… not bad, really."

"Hmm. So I guess he didn't talk much about his work."

"He did, actually. He started really early in the morning, so he was there when I came home from school. He liked tinkering with the bots, as well. Over the years, quite a few abandoned or decommissioned cleaners made it to our place. He gave them a new life." Ben grins. "He took me along sometimes—to work. On weekends and stuff. I loved it."

"I bet he loved having you along." She smiles. "I bet he was a really good Dad."

"He was. He…" Ben chokes up.

"You miss him."

"I do." Ben draws a few deep breaths.

Tilly waits.

"There was this one time," Ben continues, telling the story of the maverick bot that would not go where it was supposed to go. Halfway into the story, he realizes that this is his own story, not his father's. He does not correct it. Tilly laughs, kindly.

"Did you know him well?" Ben asks. "Back then?"

"Well, I…" Tilly looks flustered for a moment. "He was visiting from somewhere south of here. That part of the fairytale is also true. This was one of my father's many objections. He objected to Jacob because he did not know his family. Anyway, Jacob wasn't around for all that long before… before they ran off."

"Did you ever consider it? Crossing over? Joining them?"

"I never knew that they had succeeded, did I?" Her voice carries a trace of hurt again. She stops. After a moment, she continues more lightly. "Anyway, it's not exactly something you can easily…"

"But did you *want* to go?"

"I used to argue with Ella about it, when she first got the idea. She tried to convince me to go. Think of all the possibilities, she'd say. Think of never having to obey our father, ever again. We'll do what *we* want instead. She was so enthusiastic. I remember that." Tilly sighs.

"But you didn't want to go with them?"

"No. I didn't. I would have been…" She frowns, then starts again. "This is my home. I grew up in this house, in this village, with these woods and fields around. It's part of me. And there's the school."

"And Milly," Ben adds, when he feels her slipping into quiet.

"And Milly. And now you." Tilly smiles and ruffles his hair. She drains her cup and starts looking around. He knows that she will get up soon, ending the conversation. But he has one more question.

"Do you think…" he starts, thereby keeping Tilly seated. He stares hard at the much-used table between them, but does not progress.

"Do I think?" She tries.

"Do you think that the accident… that it might have been on purpose? Like if one of them was sick or something, maybe they…"

"No," Tilly says, very firmly. "Don't even think that. They loved you far too much to do that. I'm sure of it. Don't even…" She shakes her head.

Ben tries to work out how best to explain what he is thinking: Their illegal status, the risk of quarantine, possible health issues, the unexpected and unexplained fatal trip. But it is all vague and indirect. Besides, how could Tilly possibly know? So, why is he asking her?

While Ben has been thinking, Tilly has been clearing away cups and wiping the table. "Another day tomorrow," she says, gently, and touches his arm.

He gets up slowly.

Tilly bends over to pick up the sleeping Milly. This time Ben sees her wince. Pain is written all over her face. He hurries to her side.

"She has grown, hasn't she?" he says, as he picks up Milly.

They both know this is not true.

# 6

## Unrest

The chain is silver and made of irregular, flattened links. The e-stick itself is shaped like an Easter lily. Eiko likes wearing it around her neck, as a lucky charm. Before she shuts down her computer, she saves the latest version of her data onto the stick. The debilitating hack a couple of years ago has made everyone more careful about back-ups. Not that everyone necessarily thinks research data from the Reserve are worth saving at all—or that work like hers, meticulous analysis of gene expression in germline and pre-em cells from different sources, should even be performed. Scientific experimentation is what got us into this mess! Eiko refuses to give in to such defeatism. They *will* solve this problem. They have to. She pushes back her chair and gets up quickly. She waves at Janine and Liling, both working at nearby desks. Only Liling notices and returns the gesture, before focusing on her screen again. It feels wrong to be leaving the lab so early. Eiko promises herself to make up for it over the weekend.

She pauses in the middle of the lobby, as soon as she has a clear view of the front door. Much of last year it was impossible to leave through that door. There were too many angry people outside. They wanted answers. No one had any. They wanted people to blame. That was easier. The Reserve was an obvious target, with its research divisions still located on the campus of "first edits" fame. Eiko thought about moving to another lab for her PhD to avoid the politics and the protestors. But she convinced herself to wait it out. Today's students could not possibly be blamed for yesterday's mess. In fact, their work was needed more than ever. Eventually, people would realize this and stop harassing them. You are over-intellectualizing human behavior, her father said. Amita, her PhD advisor, agreed. They both said they would understand if she chose to bail from the project. She didn't. Raphael didn't exactly bail either, although he had even more reason to. He had always been more interested in trees and plants, so it was natural that he took that route. But Eiko misses having him around. Hopefully he will be at the dinner tonight.

Judging it safe, Eiko steps outside. It is a nice afternoon, cool and fresh from two days of rain and now brighter with intermittent sunshine. She decides to walk from the University to her parents' house, which will take her through Rosswold Park and along the river. She smiles at the thought. She turns the corner of the building and immediately realizes her error.

The nice day has brought out a big crowd. Today, they are clustered in front of building one, which is mainly administration. Maybe an important meeting is underway inside. She sees that there are two groups of people, a small one to the left and a larger one, center and right. She recognizes both. The smaller group has eight or ten purple-clad devotees at its center, tightly surrounded by fifty to a hundred supporters in normal clothes. Some of them are quite old, others very young. The young tend to be closer to the devotees and most of them are dancing. As she gets closer, she hears their soft-flowing voices, like a chant. The older supporters are not moving, but they are singing along. They look calm and unafraid, almost otherworldly. The larger group, now on her right, seems less organized. Near the edge, she sees some men moving about restlessly, as if spoiling for a fight. They look like hard-core members of the Charter Defense League. Others are turned away from her, apparently listening to someone giving a speech from a raised platform at the rear. She cannot hear the words, but the loud cheers suggest it is rousing. Both groups have acolytes swarming about, trying to enlist 'people's support' signatures from passer-byes. Eiko slides her right hand, where her wrist-link and ID is, under her left arm. She does not want to be counted, on either side. But she also won't be bullied into retreating. She takes a deep breath and aims for the gap between the two groups. It is daytime and a university campus, she reminds herself. Nothing bad will happen.

Moving through, the gap gets smaller. A couple of CDL types cross the divide, making brief but provocative forays into the religious group. They do not reach the devotees themselves, but move between the supporters surrounding them. The supporters, in turn, are starting to lose their calm. They cluster closer together. Blocking arms go out, some in anger. There is a line of police on either side of the crowds, ready to move in.

Eiko realizes she should have swallowed her pride and gone the long way around. She walks faster in order to slip through before the gap closes completely. Now, flash-shirts are turned on, text-mode, on both sides. She focuses on the words, in part to avoid looking directly at people's faces, in part for orientation. Texts identify. If one reads fast enough, that is. She hopes her lack of flash-text will make her invisible, or at least signal that she is neutral. To her right flashes "Our Country, Our Laws" and "No Church Tyranny", alongside "Remember the Plague, Remember HQV", "Unedited = Uneducated" and the more hopeful but somewhat incongruous "We Shall Overcome". To her left, she reads "God has Spoken" followed by "We *must* Listen", next to "The Ultimate Hubris", "Go Pure" and, surprisingly, a humorous one: "Extinction: not just for Dinosaurs". She tries to keep moving straight ahead, but finds it increasingly difficult. The gap has disappeared. Church meets Hubris. Plague meets Pure. The wearers stare at one another, with hate. Eiko sees a new text in extreme close up: "Clones are an Abomination". The woman wearing this seems to be shouting something straight into Eiko's face, the words unintelligible. Eiko freezes. The woman's face is flushed with righteous anger; her arms are raised. The flash-shirt shifts to video-mode: the famous video with rows and rows of identical clones, all saluting. Eiko unfreezes and quickly ducks to the right. "HQV *will* Kill" meets her there, a blood-red flash-text on a white background. Now this shirt switches to video-mode. It shows a young girl, convulsing violently. Yet another iconic video, this one used in history lessons. "Unedited = Uneducated" moves in, next to the HQV man. Eiko cannot get around them. She stumbles backwards instead. She imagines the righteous woman coming for her from behind, shouting.

A sudden, blaring sound arrests all movement. "The use of hyper-emotive videos has been detected. These videos are illegal at public gatherings. All IDs in the zone have been mapped. Any violence will be recorded and persecuted. Please stay calm and return to your assigned areas." The voice is loud and comes from multiple loudspeakers simultaneously. It is impossible to ignore. The order is repeated. People turn around and shuffle back to their original positions. The gap reappears. Eiko picks herself up and moves straight on, at first simply walking rapidly, without looking back, then breaking into a run. The river must be close. She can almost smell it.

\*   \*   \*

"Eiko, sweetheart," Yuriko says, giving her a quick hug. "I'm so glad you could make it." The hugs still feel awkward, but Eiko knows that their absence would be worse. Realizing how sweaty she is, she steps backward, a little too far and a little too fast, almost hitting the mirror in the tiny entryway. Yuriko steps back as well. "Are you alright?" she asks, with alarm in her voice.

"I'm fine, Mama," Eiko says, trying to calm her breath and her voice. "I ran some of the way, that's all."

"You ran?"

Mother and daughter look at each other, sifting through possible questions and answers, all unsaid. Eiko continues to breathe slowly and deliberately. Dramatic chords from a cello reach them, filling the silence.

"Papa is home." Eiko sounds relieved.

"Yes. He stayed home today. He's been cooking like a mad man." Yuriko adds a cautious smile. "Anton and Marie will be here at seven."

"OK," Eiko says after a short delay. She is listening to the cello.

"Stay overnight, won't you? Walking home at night, I always worry something might-"

"Sure." Eiko smiles. "I'll go directly to the lab tomorrow."

"Tomorrow?"

"I have so much to do." She pauses. "Is Rafi coming tonight?"

"I don't know. I'll lay a place for him, just in case."

"And François?"

"I don't think so. He never…" Yuriko shrugs and turns around, leaving the words hanging. They move into the spacious, L-shaped living room. Paul has just finished the piece and, seeing Eiko come in, starts to put the cello down.

"Don't stop, Papa," she says. "I'll join you." She walks quickly toward the grand piano, lifts the cover from the keyboard and sits down. The piano stool needs no adjusting. She starts leafing through the scores left out on the music stand. Yuriko stops in the middle of the room and studies the two of them for a moment. Eiko calls out a couple of suggestions to her father. Paul adjusts the position of the cello and tightens the bow. They settle on Brahms. They have been working on this sonata for a while. Paul leans over his instrument and leads out as Eiko lets the early chords fall softly. Yuriko glides noiselessly toward the larger of two off-white sofas. She lowers herself into the corner seat and folds her legs up, leaving the cotton slippers on the floor.

Eiko and Paul stop to confer after a passage that is still giving them some trouble. Yuriko slips into the kitchen to do the one job that still awaits her. It is the most pleasurable job, she thinks, and she likes to take her time with it. She will find just the right bowl for each of the dishes Paul has prepared and just the right plates for them to enjoy it on. Each bowl is placed gently on the counter; each plate is carried to the table on the far side of the living room. She makes hardly any noise. The music has started up again. Soon after, she slips back to her spot in the sofa.

"We're getting there," Paul says with a grateful smile to his daughter. He stands up straight, rolling his shoulders. "But now I should see to the wine."

He looks at Yuriko in the sofa. She makes as if to get up but he signals for her to stay seated. He returns the cello to its stand in the corner and hangs the bow alongside. He passes Eiko, still at the piano, and lays a hand on her shoulder. She looks up and returns his smile.

"Some Chopin?" Eiko asks her mother a moment later.

"Yes, please," Yuriko says. "That would be lovely."

Eiko looks through a large, pale blue book, finds what she wants and puts the score on the stand. She plays, however, from memory.

\*   \*   \*

"You are a magician, Paul," Marie says as she scoops the last bit of green curry from the midnight blue bowl. "Can I borrow him for a month?" She directs this at Yuriko, in a playful tone. "Or just a week?"

Yuriko shakes her head with a smile. She is glad to see Marie in a good mood. Losing the election was predictable, but the attacks on her person have been vicious. She has seen Marie Delacroix described as the most unpopular person in the country. This might very well be true. Marie still needs personal protection. Even François and Raphael were targeted for while. The threats have abated recently, allowing Marie and Anton to return to a semblance of normal life, such as tonight's dinner. Before the news broke, the two couples saw each other every few months. That was two years ago. Yuriko remembers the tumultuous summer very clearly. The phrase "fertility block", so blunt and clinical, was suddenly everywhere. The shocking facts, the monumental severity of the problem, made it seem slightly unreal at first. But soon everyone had to accept it. For once, here was a piece of news that could be confirmed by just looking around: No pregnant women in sight. Blaming the minister of science was easy. Her more extreme critics hold her responsible for the block itself, which Yuriko knows makes no sense whatsoever. Others merely blame her for the cover-up, which there must be some truth to. Yuriko had felt a sting of hurt, initially, that she and Paul only found out when everyone else did. She understands it better now, the desperate silence of the former government. The new government is not too popular, either. No one likes the choices these days. But everyone is talking about it, compulsively, urgently. Tonight is no exception.

"Do you know what is going on with the New Eden negotiations?" Paul asks Marie when they have finished complimenting his cooking. Yuriko sends him a quick, displeased look. He shrugs. "It would be weird not to ask," he says. "It's on everyone's minds."

"I don't mind you asking," Marie says. "But I only know what any other member of parliament knows. I am no longer-"

"Don't do that," Anton interjects. "We all know that political life is unfair. But Paul and Yuriko are old friends. They deserve a proper answer."

"I…" Marie looks chastised, something Yuriko has never seen before. Feeling sorry for her, she almost steps in to divert the conversation. But now that the subject has been raised, why not get some answers, she thinks and stays quiet. "I still have some contacts," Marie finally continues, "but the science ministry has very little influence these days. The negotiations are handled primarily by the minister for state security and defense."

"Defense?" Paul sounds surprised. "So we might end up with a war?"

"I hope it won't come to that. But internal security is obviously an issue— and there are some real problems along the border. Everything is so uncertain." She sighs. "Contingency plans are being drawn up for all scenarios. They must be doing the same. Preparing."

"They?"

"New Eden. They closed the transit centers last year. The centers never actually worked as intended, but symbolically… And now they are building up troops along the border. We can't ignore that, even if we believe they are doing it for internal reasons. No one has any interest in an actual invasion— not them, not us."

"Except the Charter Defense League," Anton adds.

"I suppose it's partially to appease them that the sword-rattling is going on. The real work is being done by more discreet negotiators."

"And?"

"It looks like there will be a transitional treaty soon."

"Transition to what, is the question," Anton says.

"Yes, true. But, apparently, even this first step has been really difficult. They are fanatics." Marie's mouth twitches with dismay. "They regard us, the 'impure', as evil or fallen, and, in either case, damned. They see this situation as a chance to eradicate our sinful way of life. That's not exactly an ideal basis for negotiating."

"Perhaps a slightly extreme interpretation?" Paul asks, with one eyebrow raised.

"Not really." Marie shrugs. "It's how I've always seen it. Now that I'm no longer in charge, I don't have to be quite so diplomatic about it." She sighs. "We had to make huge concessions just to get them to sit down with us. Just to talk. For example, practically anything to do with youngtwins had to be stopped immediately. I'm not sure why, but they are particularly sensitive about this. Maybe it's historical." Yuriko and Paul exchange a look. Eiko sees it, but Marie does not. Unless Raphael has told them, they won't know about her. Marie continues. "It looks like most of the regular work at the Reserve will be disallowed by the treaty, not just suspended."

"Not that it makes much of a difference," Yuriko says. "Sadly."

"Perhaps not in the clinics," Paul says, somewhat under his breath. "But this could affect the research sections, as well. We are trying to-"

"It's not just the Reserve they're after," Marie goes on. "They want to ban active end-of-life decisions—not surprising, but still… and possibly somatic editing, as well. The requests have already been floated."

"Not *all* somatic editing?" Yuriko exclaims. Marie nods. "But that would be terrible! So much of our cancer treatment depends on therapeutic response editing. Our negotiators won't give in on this, will they?" Yuriko implores. "It would be… devastating."

"I hope it doesn't come to that," Marie says. "But I doubt Gareth understands how significant somatic editing is to our health effort."

"Gareth?"

"The new minister of science, Gareth Svensson. Apart from having essentially no science background and no medical background, his hearing is also severely defective. He only hears those who shout."

"Perfect for the job," Anton says, with sour sarcasm. "He still hasn't had time to meet with University representatives. He's too busy with the so-called 'concerned citizens' groups."

"Well, given the general mood of the public, I can understand why he feels the need to pay attention to them," Yuriko says, carefully. "At least for now."

"If the public could only understand the consequences of endlessly giving in to these holy blackmailers. We're destroying our culture," Anton says. "I do get where the CDL is coming from, even if they can be a bit crude."

"A *bit* crude?" Marie says, angrily. "The CDL has suggested we use force to get fertile mates! They make us look like monsters." She glares at Anton. "*And* they are making everything worse internally by deliberately provoking the religious groups." Eiko remembers her afternoon experience and shudders, but remains quiet.

"Well, the New Eden priests think we are monsters no matter what," Anton mumbles. "So why not act the part?"

"Anton, you don't mean that!" Marie says, severely.

"No, I don't," Anton admits. "I'm no hawk. Too much of a wimp." He smiles at the others, sheepishly. "Of course we shouldn't use force—or, if so, only to open the borders. Maybe we shouldn't have built that damn wall to begin with." His smile widens. "How about that? We open the borders and give a warm welcome to any unedited youngsters who wish to come here and agree live under chartered rules. There'd be plenty of them, I reckon." He turns to Marie. "You told me the dome population was growing rapidly until New Eden sealed the borders."

"Let's not talk about the domes," Marie says, peevishly. The rapid expansion of the domes was the most visible, and most damning, evidence against the former government. Marie remembers the argument all too clearly. The upper levels of government must have known about the fertility problem for quite some time. How else could they explain the dramatic increase in facilities designed to allow non-charter, unedited refugees to settle? They couldn't. They *did* know.

"Unfortunately, we can't just let unedited people move here," Yuriko says. "Until we have more information, we need to keep them separate somehow."

"I suppose I knew that," Anton says. "Risk of infection."

"Bringing in HQV-infected individuals is a risk, of course," Yuriko continues. "But we could prevent that relatively easily by temporary quarantine and screening. The question is more what happens down the line. Unedited people will remain susceptible to HQV, as will any children they have, even with an edited partner. Only fully edited people are resistant."

"Hybrid offspring may pose even more serious problems," Paul adds.

"Hybrid offspring?" Marie says to Paul with an expression of disapproval. "You make it sound like we are lab animals in a biological experiment."

"That's not completely inaccurate, actually," Paul starts, "but what I mean is…"

"What he means is," Yuriko says, "that hybridization between edited and unedited genomes, as you'd get in mixed-parent couples, creates a new and potentially dangerous situation. Hybrids can be infected by HQV and, once infected, they might allow the virus to pick up our genetic changes and thus jump the resistance barrier. It's entirely possible. And, if that happens, we would have the crisis all over again. Plus more anger and more blame."

"So there's no advantage to… a little mixing?" Anton asks.

"Temporarily, yes," Yuriko says. "But in the long run? It could end up a nightmare." She frowns. "Another possibility is that we allow mixed couples to have children via IVF and put the edits back in, immediately. Of course, we don't know for sure whether mixed couples avoid the fertility block. And, in any case, the edited children would presumably be subject to it."

"Eventually," Paul muses. "But we would buy ourselves another generation—thirty, forty years—to solve it in." Yuriko looks at him and nods slowly. Eiko does the same. Marie looks at him with a strange intensity. Paul wonders why.

After a while, Marie clears her throat. "We actually have a substantial mixed parentage population, or hybrids as I guess you'd call them, in the domes already." All eyes turn to her in surprise. She purses her lips and continues. "I shouldn't be keeping you from talking about the domes. They are not exactly secret any more, are they?" She sighs. "The old domes were quarantine stations, built to handle a small number of motivated people for a limited time."

"Like Doctor Vera Weiss," Eiko says, excitedly. She immediately regrets her outburst, but no one seems to find it strange.

"Yes, people like her," Marie continues. "The new domes are very different. They are essentially self-sufficient communities, with small-scale agriculture, schools, health clinics and so on. The idea is that people can live well there—for a long time."

"They might have to."

"Yes—but the domes are designed for it. They are spacious and pleasant, not overpopulated refugee camps. The news outlets are being irresponsible." She adds a quick huff, in irritation. "Not a single image they have shown is from inside one of the new domes."

"They couldn't get in, could they?" Paul says.

"Of course not. You can't just have random photographers running in and out…"

"But why not transparent walls?"

"I know, I know. We should have realized that transparency—literally—would have been a better choice, but we…"

"Didn't want to acknowledge why the domes were needed."

"Well… Anyway, this is all being corrected now. The domes will soon be open to the public through large viewing and interactions areas. There'll also be live streaming—in fact, one or two of them are already online, as of last week."

"So we can watch the poor refugees like animals in a Zoo?" Anton says. "That doesn't seem very-"

"I don't like the idea any better than you do, Anton," Marie continues, "but it was not my decision. It wasn't even discussed in the public health committee. We were simply told about it two weeks ago." Anton sends her a quick look of reproach. She brushes it off. "The hope is that people will be soothed by seeing the newborn babies and watching them grow, seeing that new life is still possible."

"And what about the CDL?" Anton asks, with irritation. "Won't a display like this provoke them? Fuel their desire for action?"

"I don't know." Marie stares him down. "I hope not. But that's the current plan. There are already quite a few babies and young children, as well as pregnant women, in the domes. Women from New Eden are given a choice: They can have unedited babies, as they would back home. They just need to be registered and, for now, stay in the dome. We don't require somatics any more, just a clean bill of health and continuous reproductive monitoring. Of the child as well, eventually."

"You mentioned hybrids," Paul says. "Individuals with mixed parentage. Are they children or adults?"

"All ages, actually," Marie says. "They generally come from the border area."

"I've heard a rumor that if the mother is unedited, there's no fertility problem—whoever the father is," Paul says.

"That's what I've been told as well." Marie nods.

"And the other way around?" Paul asks.

"Interesting question. I don't know." Marie tilts her head. "Any useful rumors?"

"Not that I know of."

"Anyway, they came forward voluntarily. Border communities were contacted and non-registered people, whatever their status, were guaranteed citizen status with no repercussions, no forced somatics, reproductive freedom and, if needed, a safe home."

"*If needed* meaning everyone who is susceptible to HQV?" Yuriko asks.

"That's my understanding. Quite a few came out of hiding."

"For a home in a dome." Anton says, drily.

"It was considered the safest option," Marie responds, curtly. "It still is. And the domes *are* very large."

"You said the New Eden women are given a choice," Yuriko says.

"Yes, those willing to have edited babies have been encouraged to do so. They have been given all the support we can offer: free health care, free education, good housing. We call them pure spring babies." Marie smiles briefly. "Pure mothers, edited offspring. Fathers may be edited or unedited. As we've just discussed, both combinations seem to work."

"But these pure spring babies," Yuriko says, "they may well be subject to the block when they grow up."

"Possibly. Probably. But, like Paul suggested earlier, it buys us time."

"So all this has been in the works for years?"

"The public spectacle is new, but the rest, yes… a few years. The required editing has been done at the Reserve, discreetly." Marie sighs. "Of course, we may soon have to stop this. The New Eden negotiators know about our pure spring babies and insist we stop the practice immediately."

They all fall quiet.

It is Anton who breaks the silence. "Obviously, the negotiations are leading our country down a terrible path," he starts, addressing Yuriko and Paul. "We desperately need to find a real solution. So, what's the word?" He looks back and forth between them. "*Can* we beat it?"

"Well," Paul starts, reluctantly, "we're doing our best. Had we known about it earlier," here he looks pointedly at Marie, "we might be further along by now."

"Everything the ministry did and did not do was after consultation with the leadership of the Reserve," Marie says, acerbically. "They knew as much as we did. How many of the scientists they chose to share it with was their decision. We were all worried that taking it public would lead to panic and civil unrest."

"Which is exactly what we've got now," Paul says, irritably, "but with the added benefit of complete mistrust in the government. People have been lied to before, but rarely about something this important and this close to the heart."

"It was a political decision. And—I admit—maybe not the best one."

"If we had been told earlier, we could have-" Paul starts again.

"I hear what you're saying," Marie interrupts him, wearily, "but the leadership *did* know-"

"Twenty/twenty hindsight," Anton interjects. "How about we get back to the actual question here?" He raises a forefinger in Paul's direction. "You said the scientists at the Reserve are doing their best. But where are you at? *Will* there be a solution? Soon? Ever?" He lowers his finger. "Sorry. I'm not trying to put you on the spot here. I-" he glances at Marie who nods in agreement, "*we* just want to know your honest, professional opinion."

"During the crisis, scientists came up with creative, effective solutions. So I believe we *can* do it," Paul says slowly and deliberately. "We just have to keep trying."

Eiko cringes in her seat, willing her ever-cautious father to step up.

"But is there any actual progress?" Anton continues.

"We don't have the solution yet. We don't even have the outline. That is the honest truth—unfortunately."

"But we *do* know something," Eiko says, speaking rapidly. All eyes turn to her. She blushes. "We know the block acts at the implantation stage, probably after zygotic transcription has started. For earlier stages, including germ cells, cell behaviors and expression profiles look normal." She pauses. "It's a good start."

"Why is that good?" Marie asks in an encouraging tone.

"We know how to work with pre-em cells. If the problem is in the early embryo, we are better able to do something to bypass it."

"Hmm." Anton nods. "That sounds promising, I guess." He looks from Eiko to her parents. "And is there a working hypothesis as to what exactly the problem might be?"

"Well," Paul and Yuriko start at the same time. Yuriko gestures for Paul to continue. He does. "Some facts are clear: all edited populations, which means all chartered countries, are affected. None of the unedited are." He stops.

"*However*," Yuriko takes over, "the only edits we all have in common are for HQV resistance and these have been around, and homozygous, for several generations. You know about the original editees?"

"Of course," Anton says. "They were born during the crisis, so about a hundred years ago. Are they still alive?"

"One of the five was killed by a terrorist attack during the crisis," Marie says. "Of the remaining four, one died a few years ago, I think."

"So good long lives," Paul adds.

"*And*," Yuriko continues, "they have healthy children and grandchildren. So, obviously, the HQV edits as such are not the problem. IVF procedures are not, either. Once everyone was homozygous for HQV resistance, mandatory IVF was abandoned. That was twenty years ago. Some people have chosen to conceive the old-fashioned way since then." She stops and adds a quick smile. "Unfortunately, the block still applies."

"Too bad," Anton says, smiling as well.

"So, discounting a bolt of God's wrath," Paul takes over, "we have to conclude that the block is caused by intersection."

"And by that you mean… what?" Anton asks.

"The agent directly responsible for the block must have arisen or spread recently. It interacts with, or depends upon, the HQV resistance edits, since it does not affect non-edited genomes. So—an intersection of conditions, an unlucky coincidence."

"So, you're saying that the original edits may have taken us out of the frying pan and into the fire—that just wasn't lit yet?" Marie asks.

"Yes, you could say that."

"I revisited the old videos," Marie says, "the recordings of individuals with full-blown HQV-infection. It's really horrible. We are *not* going back to that."

"I saw one of those videos today," Eiko says. "On a flash-shirt."

"Really?" Marie says and frowns. "But it's illegal to show that footage in public. It's considered highly incendiary."

"The police did step in," Eiko says. "It was a big protest, on campus. Both sides. It got a bit…"

"What happened? You didn't say-" Yuriko starts.

"I wasn't hurt or anything." Eiko sends her mother a quick, reassuring smile before looking around the table. "But what if the virus is no longer around?" she says. Eagerness has crept into her voice. "That would change everything, wouldn't it?"

"Well, yes. But…"

"I've heard that it might be. Gone, I mean."

"Did you hear that from those purple-robed idiots?" Marie exclaims. Yuriko sends her a displeased look. "They *are*. I've met them," Marie continues. "Naïve fools. They know nothing about New Eden."

"No," Eiko says, more quietly. "It wasn't from them." She pauses. "It was from a New Eden priest, Father Elias, two years ago—when we were at the transit center."

"Well, they may—and they do—claim that HQV has been eradicated," Marie says, with a sigh. "But we have no way of knowing whether it's true. They won't let our teams in to investigate. That *does* suggest…"

Yuriko and Paul both nod a few times.

"So, you want to go for the direct agent?" Anton asks Paul. "According to your intersection hypothesis, there should be one."

"Yes. We all think that's the way to go. It's just…"

"We don't know what it is," Yuriko says.

"Argh," Anton exclaims. "Really? No idea?"

"We are following up on several possibilities. It could be a virus with opposite host preference to HQV. Or perhaps some other bug."

"It would have to be totally asymptomatic—other than the sterility, that is," Yuriko says. "We should have identified something like that by now, or traces of it. But… If it's very unusual, maybe not. We'll keep trying."

"Which gets us to the second option," Paul continues with a hint of excitement, "and this is what I'm betting on: A prion-like agent."

"Prion?" Anton asks.

"It's infectious, which could explain the spread. Lots of people travel within and between chartered countries, all the time. But a prion is not based on nucleic acids, so it's much harder to identify—unless you know exactly what you are looking for."

"Hmm. Is there a third option?"

"Something chemical, non-biological. But it would have to be new *and* be everywhere *and* have this very specific biological effect… Hard to imagine."

"In each of these scenarios, you postulate that something attacks and screws up our cells, but doesn't touch theirs," Anton says. "Is that even possible?"

"It *is* possible," Paul says. "The HQV edits change two interacting cell surface proteins found in nerve cells, immune cells and germ cells. So our cells *are* a bit different."

"Which means you just need to find this pernicious agent."

"Yes."

No one says anything for a little while.

"But, if you *don't* figure it out," Marie says, "our future and, in fact, the future of humanity," she pauses for emphasis, "depends on New Eden and other non-chartered territories. And that means we have to work with them."

"I thought you…" Anton starts, looking at his wife with surprise.

"I have my preferences, sure I do. But we need a back-up plan in case science does not come to the rescue this time. The occasional refugee is not going to be enough. So, someone has to negotiate with New Eden." She sighs. "I suppose I'm glad it isn't me. Can you imagine making a political decision that weighs future children against treatment of current cancer patients—for everyone?"

"Whoever does it is certainly not re-electable," Anton says. The others nod.

"A minor sacrifice," Marie says. More nods.

"So, we're back to the negotiations with New Eden. Is there some other way forward?" Paul asks, looking at Marie. "Other than more restrictions, I mean. Is there anything they want that we can give?"

"Unfortunately, that's a difficult one. Their population is not as dense as ours and they are pretty much self-sufficient."

"Medicine?"

"That's one piece of leverage we *do* have. Editing is a sin, but they don't mind using antibiotics and other drugs. They produce some of it themselves, but we've been exporting the more advanced stuff to them for quite a while. This is being strategically delayed now, but we don't know about stockpiles."

"There's also the imbalance in weapons and defense technology, especially shields," Anton adds. "The science behind it is common knowledge, but effective commercial versions have all been developed here, mainly by Huang Shields. The guys in New Eden eventually get their own, but as long as we stay one step ahead…"

"I don't trust Huang," Marie says.

"Me neither," Anton says. "Victor Huang is a dangerous operator. He already has way too much power; it'll only get worse. He argues that his company needs to be able to move more swiftly—unimpeded by regulations. In the interest of national security, of course."

"We never-" Marie starts.

"I know you resisted that," Anton says. "But the new government? Are they going to be as conscientious? When push comes to shove?"

"Does New Eden buy from Huang Shields?" Eiko asks, trying to remember what she once heard and from whom.

"Not exactly," Anton says. "They mostly copy the products. They also-"

"But Victor Huang is an oldtwin," Eiko says quickly. "I've met his son, Leo. I'm sure he's Victor Huang's youngtwin."

"Ha!" Anton shouts. "All that hatred of youngtwins and it turns out they are actually *copying* defense technology from one of the culprits. Fantastic!" He smiles and shakes his head, clearly loving the irony of it.

"How do you know the Huangs?" Yuriko asks Eiko, sounding worried.

"I don't *know* them, not really. Leo was part of the group I went to the transit center with—a friend of Ben's."

"So they know?" Anton asks, excitedly. "New Eden knows about Victor Huang and his youngtwin?"

"Yes," Eiko answers, after thinking for a moment. "They must know. That was the reason Leo wasn't allowed in."

"Marie! Did you hear that? This is perfect!" Anton says. "It's a solid, objective reason to keep Victor Huang from playing a role in the negotiations."

"Maybe." Marie frowns. "But politics and business are pragmatic endeavors. Principles get ditched if they become too inconvenient."

"But if everyone *knows* that Victor Huang has a youngtwin, it would be impossible for the New Eden side to engage officially with him. They are too fanatic about this issue."

"I have no say in this, not anymore."

"But you could let the information slip." Anton keeps looking at his wife. "Someone would pick it up." She does not answer.

Paul turns and looks at Eiko. They eyes meet. She wonders whether he has just worked out what she did not quite say. That *she* hadn't been allowed in. They never discussed that part, she remembers. She hadn't said. They hadn't asked. "That crazy trip of yours," Paul says now, his voice strained. "You should have told us you were going. We tried to trace you after a couple of days. It was a nightmare."

"So did we," Marie says, turning to Paul. "I even got the police to…" She stops. "Rafi wouldn't tell us anything about it."

"We weren't in any danger," Eiko says, firmly. "They treated us very nicely—and with respect." She hadn't told them that part either, had she? She hadn't mentioned how much she had liked Catherine. They had connected, somehow. "It was only Leo who acted… strange," she continues, "like he had something to prove, or it was some sort of game to him. And then, after Ben…" She goes quiet.

"Did you ever hear from Ben again?" Marie asks quietly.

Eiko does not respond.

"No, we haven't heard from him," Yuriko answers for her.

"I did everything I could to track him down afterwards—through diplomatic channels," Marie says, directed at Eiko. Eiko remains quiet. Marie addresses the rest of the table. "And then all hell broke loose."

"We know you did," Yuriko says. "And we appreciated it. We were very concerned about Ben, but…"

"Ben was unedited," Eiko says. "Is. He *is* unedited. It wasn't just his parents, like we thought. That was one of the reasons he crossed over." A touch of anger enters her voice. "He wasn't allowed-"

"No," Yuriko says, firmly. "That's not possible. Ben went to school here. So he couldn't be."

"But-"

"The rules are very strict," Marie joins in. "A child can't start in a chartered school without a clean Med-data file. If the file had shown him as unedited, he would have been-"

"It didn't," Eiko cuts her off. "His Med-data file was normal. It must have been…" She stops, a frown on her face.

"But if the file… what makes you think Ben was unedited?" Marie asks.

"Father Elias," Eiko says, her voice no longer confident. "At the transit center." She had believed it, back then. She tries to remember why.

"The Father Elias who also claims that HQV has been eradicated from New Eden?" Marie asks, pointedly.

"Yes," Eiko admits.

"Well, then!" Marie adds a nod of finality. Yuriko sends her a quick look. Marie turns back to Eiko and softens her tone. "I know you and Ben were close. Rafi said… well, I wish he had told us more about that trip."

"The trip was for Ben. He needed answers," Eiko says, sounding hurt. "The priests must have helped him cross over. Why would they do that unless he was one of them?"

No one answers.

"You know what?" Marie suddenly lights up. "You should check out the domes. He might actually be back. We still had a few come across last year. If Ben really *is* unedited, as you say he is, he would have to go to a dome on his return. Even if he is a citizen."

"And he would still be in there?"

"It's certainly worth a try." Marie purses her lips. "I know some people who can help. Let's find out."

Marie pulls up the names and locations of three domes on her wrist-link. She transfers the data to Eiko's link and leans toward her to add various bits of information about each of the places.

Yuriko starts clearing the dishes. "Dessert anyone?" she asks.

"Yes, please," comes from Marie and Eiko in unison. They move back from their close exchange, both smiling.

"How is Rafi, by the way?" Eiko asks. "I haven't seen him for ages."

"Busy," Marie says with an unhappy half-smile.

Anton supplies a non-committal shrug.

During dessert, the conversation returns once more to politics and the muddle of the negotiations. Eiko's mind is elsewhere. After a while, she leans over to her mother and whispers in her ear. Yuriko nods with a smile and whispers something back. She reinforces the smile when her eyes catch Paul's. Eiko slips quietly off her chair. Soon one of Beethoven's sonatas floats gently from the other end of the room.

*   *   *

Eiko has shifted back to Chopin, the prelude in B minor. Marie and Anton have just left, with repeated thanks to Paul and Yuriko and with promises to reciprocate soon. Before leaving, Marie walked over to Eiko at the piano, gave her a quick hug and whispered in her ear. "I'll get you that dome access first thing. I promise." Then Anton signaled from the doorway and Marie hurried to join him. Once they were out the door, Eiko started playing again.

The prelude finishes. Eiko does not start anything new, but stays at the piano, her eyes on the keyboard. Yuriko comes into the room from the kitchen. Seeing Eiko so still, she stops moving as well. She waits.

"Mama?" Eiko finally looks up.

"Yes, sweetheart." Yuriko moves swiftly across the room and sits down on a small stool close to the piano.

"It would have been her birthday today, wouldn't it?"

Yuriko looks startled. "Yes, but…"

"Nariko would have been… thirty-one, wouldn't she?"

"Yes," Yuriko answers, solemnly. Eiko does not add anything so Yuriko feels she must. "How did you know? Her birthday, I mean?"

"You told me. Two years ago. When you and Papa finally…" Eiko looks away, her voice unsteady. "And I never forget dates."

The simple statements sit there, for a while.

"I am so very sorry, Eiko," Yuriko finally says. "I know we should have told you earlier. But we…"

"She played the piano as well, didn't she?"

"Yes," Yuriko answers, hesitantly. "She started very early. It was obvious that she had a talent for music. Just like your father and…"

"Yes. Of course." Eiko's voice is neutral, but bordering on hard.

Yuriko's eyes threaten to water. She almost gets up to leave. But she knows it would be the coward's way out. So she stays seated.

"Yes, *damn right* you should have told me." Eiko's voice is suddenly raw with emotion. "I shouldn't have found out at my reading, from a state-appointed councilor who just assumed…"

"No, you shouldn't have. It was wrong of us." Yuriko pauses. She continues very softly. "When you found out about Nariko, I remember you stopped playing for a while. We assumed it was because you wanted to… choose for yourself. So we didn't push you. You moved out, got your own place." She sighs. "A few months later, when you started coming home again and started playing the piano again, your father and I were so happy, so relieved. The sun shone again. This room came alive again." She smiles. "At that time, we had another talk, a proper talk, about Nariko. You said that you weren't angry anymore." She looks carefully at Eiko. "Has something changed?"

"No, not really."

"*Are* you still angry with us?"

"No, I…"

"Did something happen today?" She pauses. No answer. "Or was it all that talk about the block and the New Eden demands? I'm afraid it's all we think about these days."

"It's what we *all* think about," Eiko says with a touch of irritation.

"Of course, of course," Yuriko says, hurriedly. "Eiko, sweetheart, you know that you mean everything to us, don't you? Because of who you are. Not because-"

"But, tell me again," Eiko says. "Why did you choose to have a youngtwin? You could just have had another baby, a regular baby."

"No other baby could have been more wonderful than you. We have always loved you and cherished you."

"I know, but… Now I am—this." Eiko frowns.

"You are a twin. There is nothing wrong with that."

"But I'm not. I'm just me, alone, and I…" Eiko thinks for a bit, concentrating hard. "Sometimes I feel like I have this extra shadow, but always in front of me. Sometimes it feels like a big black hole. It scares me."

"Sweetheart …"

"Sometimes it's the opposite, sort of." Eiko smiles, briefly. "It's like she keeps me company." She looks at her mother. "She talks to me. She gives me advice, tells me to stand up for myself. Like a big sister." Yuriko looks concerned. "Oh, no, don't worry, it's not an actual voice," Eiko adds quickly. "It's just in my head. It's a feeling." She sighs. "Sometimes, I feel so impossibly different from everyone else. They have real siblings. I'm just…"

"Oh, sweetheart. Everyone feels different."

"You didn't answer my question."

"What?"

"Why. Why did you request a youngtwin?"

Yuriko looks at her hands, firmly clasped in her lap. "Maybe it was the wrong thing to do. I don't know." She pauses. "When Nariko died, I simply collapsed. Later, when I tried to face the world again, I got lost, lost in a fog. I think I told you some of that, two years ago." Eiko nods but does not speak. "The youngtwin idea was my only light. It became an obsession, I suppose. I couldn't let it go. Even the word sounded so full of hope: Youngtwin." She smiles. "I realized I might get to see that precious expression once again, that special way you had of frowning when you had to concentrate—when you grasped a toy with your tiny fingers and later, when you started to walk. It was all I could think of." She sighs.

Eiko waits. She has heard this before. She is aware of that. But for some reason she needs to hear it again.

"I had to hide my desperation," Yuriko continues, "and make the decision look rational to the councilors. We dutifully listed all Nariko's unique and wonderful traits, musical talent included. But the truth is, I was in a fog and could see only one light. I understand that now. I was selfish. We both were. Paul…" She stops, starts again. "When you were born, we were just so incredibly grateful, so happy. You were perfect. You have been—you are—very much loved, Eiko."

"I know."

"As yourself, not a substitute." She looks into Eiko's face. "That's why we never felt the need to tell you. You and Nariko were different, as twins are. Even when you were little, you would do these…" Yuriko closes her eyes and shakes her head slowly at her memories.

"I know," Eiko repeats. "And it's OK now. It really is." She looks at her mother's hands, still clasped in her lap. She leans over and puts her right hand on top them, just for a moment. "It was just a wobble. That's all. It's been a strange day."

They sit for a while in silence.

Eiko starts looking through the pale blue book again. She stops at a page, moves her head slightly as if she can hear the piece and then places the book on the stand, ready to continue.

Yuriko has risen from the stool and stands with one hand on the piano. "We could move the piano to your place," she says. "I don't play anymore. I know it would be a tight fit, but that way-"

"No. The piano should stay here," Eiko says. "I prefer playing it here."

"My precious girl." Yuriko kisses Eiko atop the head and moves off toward the kitchen. Her face shows relief mixed with exhaustion.

Music fills the room again.

\* \* \*

"Where are you girls off to this morning?" Yuriko asks as Eiko gets into Celia's pod.

"Just a little zip around town," Eiko says. "Then the lab. I'll call you when I get home tonight."

"Alright. Be careful."

"We will," Celia and Eiko respond in cheery unison. They close the pod doors and Celia takes off swiftly.

"Will you put the coordinates in?" Celia asks.

Eiko transfers the data from her wrist-link, while Celia maneuvers around a slower stream.

"Why didn't you tell your mother where we're going?" Celia asks. "It's not super secret and terribly dangerous, is it?" She sounds like she hopes it is.

"They don't need to know everything," Eiko says, a bit primly. "Plus, I'm sure they'd worry—although there is absolutely no reason to," she adds with a sigh. "Marie Delacroix was the one who arranged the access and gave me the coordinates." She turns to face Celia. "They came around for dinner last night."

"So her house arrest is finally lifted?"

"Apparently, it was never house arrest. It was for her own safety."

"I suppose that makes sense." Celia makes a face. "And how was our distinguished former minister? Still trying to defend her terrible decisions?"

"She—she's not a bad person. And she didn't make the decision on her own."

"I know. But she could have... Well, never mind. I shouldn't complain. I got my starting capital from that snafu." She opens her eyes wide in pretend disbelief. "Shorting based on the words of a New Eden priest. Wild. You can't call it insider trading, that's for sure."

"Not that money *really* matters with possible extinction looming over us."

"Of course not!" Celia exclaims. They both smile at their usual exchange. "I trust that you guys will figure it out—you, your father and the country's entire collection of intrepid scientific wizards."

"You are an irredeemable optimist."

"Absolutely. The HQV crisis was overcome. Science can do it again."

"Mankind *will* eventually kill itself off."

"Nah… All these years of nuclear bombs powerful enough to blast the planet to smithereens—and we are still here."

"Eventually, our luck will run out."

"Eventually, you will tell me what is really bothering you." Eiko does not answer. Celia continues, more gently. "Last night, when you called, I could tell something was wrong. Why do you think I agreed to go on this crazy mission?"

Eiko frowns, seriously.

"I don't mean crazy *that* way," Celia continues. "It would be wonderful if we could find him." She pauses. "But it's not terribly likely, is it?" Eiko shrugs. "It's just… I'm sure he would have contacted you if he were back. The domes do allow communication." This gets another shrug from Eiko. Celia waits another moment, then shifts to a brisker tone. "Was Rafi there? Last night, I mean."

"No. His parents seem to see him even less than I do."

"Hmm."

"Have you seen him recently?"

"A few weeks ago. I call him up whenever I get desperate." Celia catches Eiko's disapproving expression. "I didn't mean… Eiko, that was a joke." Eiko doesn't respond. Celia continues a few moments later. "I will always love Rafi. You know that. We're just in such different worlds now. Plus he's gotten more and more…" She doesn't finish the thought. Eiko doesn't ask. "I thought you ran into him on campus, sometimes."

"I used to."

They go quiet.

"I learned something else last night," Eiko says with an inviting sing-song.

"What?" Celia smiles.

"Do you remember Ben's friend Leo, the guy who went with us to—who brought us to—the New Eden transit center?"

"Sure. Intense guy."

"So, Leo's father is Victor Huang." Eiko tells Celia what was said about Victor Huang at dinner. They talk about the Huangs for a while, then about Huang Shields and New Eden making copies. Celia echoes Anton's pleasure in the irony of the situation. Eventually, they get back to the trip to the transit center.

"We had no idea what we were getting ourselves into," Celia says, shaking her head. "What were we thinking?"

"That Ben would get some answers."

"Well, we *all* got some of those." Celia sighs, looking straight ahead.

They remain quiet for the rest of the trip.

*   *   *

"Ben Hatton, you said? Twenty-two years old?" The reception area looks like it is still under construction. But at least the desk is manned. The woman behind it did not seem surprised to see them, either. Forewarned, maybe. "And you are?" she continues.

"My name is Eiko Carr. I'm a PhD student at the Reserve. I was informed by Doctor Chen from the Ministry of Science, Department of Genetic Resources that I should ask here." Eiko scans her wrist-link to show the authorized access note that Marie forwarded to her this morning. She must have been up early, Eiko realizes.

Celia says nothing.

The receptionist taps on her desk screen a couple of times, then looks up. "Sorry, we don't have anyone here by that name."

"How about the other domes? Could you check?"

"This is the central register." The receptionist seems slightly offended. "I *did* check."

"Sorry, I…" Eiko thinks. "Could he be here under a different name?"

The woman looks at Eiko and waits. "I'd need something more specific than male, aged twenty-two," she finally says.

"Of course." Eiko pulls out an e-stick. "His sequence file is here—under 'Ben'." She hands the stick to the receptionist. "There may be a couple of errors," she adds.

"No matter." The woman attaches the stick to her desk screen and gets busy. In the meantime, Celia has turned toward Eiko with a big question mark on her face. Eiko mouths "I'll explain later."

"No, sorry," the woman says. "He's not here. With this," she nods at the e-stick while handing it back to Eiko, "we can be quite sure."

"That's too bad." Eiko looks more relieved than upset, Celia notices. "Thanks for your time," Eiko continues. She looks to the left, toward a glassed-in corridor. "Is this the public access area?"

"It will be. It's not… complete."

"Can we have a look?"

"Well, there is no one… Sure." The receptionist flashes a smile of unadulterated professional courtesy.

"Ben's sequence file?" Celia says as they walk down a short, empty corridor. "Why do you have Ben's file?"

"Coincidence, really," Eiko says. "After the reading, Ben was curious about a potential relative. You know his parents…"

"Yes, of course. But why you?"

"Doctor Vera Weiss. Does that name mean anything to you?"

"No. Should it?"

"I suppose not. It's complicated." Eiko frowns. "She lived in a facility like this one for a while—many years ago." They pass the second pair of double doors and enter what looks like an enormous, many-facetted greenhouse. This is the structure they had noticed from the outside. "Maybe not exactly like this one," she adds, "maybe a bit more primitive."

"Wow," Celia says. "It's huge. And nice, actually."

The large, open area is green and luscious. One half looks ornamental, park-like, with bushes, flowerbeds, patches of grass and scattered benches. There is even a small pond. The other half, further along, looks like a very large vegetable plot or a field of mixed crops. They walk along slowly, taking it all in. They come close to a bright orange, flowering bush and Celia reaches out to touch it. She is stopped by a transparent wall and laughs in surprise. The corridor they are in runs all the way along one side of the dome. The other side has an array of opaque double-doors, which appear to connect to the low buildings behind the dome.

But there are no people.

"Well," Eiko says, looking disconcerted, "I guess she was right. It is not complete."

# 7

# The Witch

SCHOOL IS STUPID. I WILL NOT GO.

Ben looks from the note in his hand to Milly's face. Under the hard fringe that Tilly cut for her first day, Milly's brown eyes look straight back at him. Her mouth is pursed in fierce determination. He feels like laughing at the pluck of it.

"It's hard—in the beginning." He puts a hand on her head. "I understand that. But school is important, Milly." Her expression remains unchanged, so he continues gently. "I'm afraid you have to go."

She looks at him for while longer. Then she goes back to the long table, picks up the red crayon from before and finds another scrap of paper. She concentrates hard while printing something on it. She hands the note to Ben.

I CAN READ. I CAN WRITE.

"I know you can," Ben says. "And I know Tilly is very proud of you. But you have to do as she says, OK?" For the past few weeks he has been helping out at the school, so he knows what the real problem is. He knows who the bullies are. He would prefer to teach Milly at home, in the evening perhaps, but Tilly is adamant that she go to school with the others. He looks toward the back corridor and sees that the door to his room is open. *Their* room, these days: It now contains a second, small bed as well as a few toys and a colorful, well-stocked low bookcase. "Go find something for us to read tonight. We need to start on a new book." Milly wavers, but does not move, not immediately. Ben turns around to continue preparing greens for the evening salad. The vegetable plot is overflowing this year. He picks up one of the crooked carrots and marvels at the amount of soil it has hung on to. Maybe that's why they taste so good. Behind him Milly is stomping out of the kitchen. The fact

© Springer Nature Switzerland AG 2020
P. Rørth, *The Unedited*, Science and Fiction, https://doi.org/10.1007/978-3-030-34624-9_7

that he can hear her means she wants him to know what she is doing. He smiles but does not turn around.

When Milly comes back, she goes straight up to Ben and pulls on his shirt. She is not carrying a book and does not bother with a note. She looks very upset. She pulls on his shirt again, harder. He puts down the knife and follows her back down the corridor, where he hears a faint moaning. It seems to be coming from Tilly's room. He knocks on the half-open door. Tilly is sitting on the edge of her bed, but completely bent over. Ben rushes across the room and crouches down in front of her.

"I'm fine, I'm fine," Tilly insists. But she lets Ben take hold of her unsteady hands. "It's just…" She looks at Milly, then back at Ben, pleading.

"Milly, could you go pick some of the very small tomatoes for the salad, please?" he says, knowing how much she loves them. "I forgot."

Milly looks from Ben to Tilly and back again.

"Do as Ben says, pumpkin," Tilly says.

Milly looks at her, still hesitating. Her lips are quivering slightly. Finally she turns around and hurries out the door. They hear her quick little footsteps disappearing down the corridor. They hear the kitchen door close.

"Is it bad?" Ben says. "Is there anything you need?"

Tilly closes her eyes and nods. The pain must have gotten a lot worse, he thinks, starting to panic. Not knowing what else to do, he sits down on the bed next to her.

"You must go to Margot," she says, when she is finally able to speak again.

"Margot? The witch?"

"So you've heard that?" She almost smiles, but then winces again. "Tell her. She has herbs for the pain."

"You don't want me to get the doctor?"

"No. Just Margot."

"But how do I find her?"

"Ask Father Len."

"Father Len? How does he know…?"

Tilly does not answer him. She breathes deeply and puts her left hand on his right arm.

How can everything change so much, so quickly? he thinks. What will I do if… He feels the panic dig further at his insides. He dare not look at Tilly. Instead, he looks straight ahead, at the open door. The panic finally eases.

"I'll go find Father Len now," he says, his voice almost normal. She shakes her head. "After dinner?" he tries. She nods. It will still be light enough, he thinks. She starts to lean sideways, away from him, her eyes closed again. He helps her lie down and goes back into the kitchen. He had managed to not think about Tilly's illness for almost an hour. Now the tears are making it hard for him to do anything at all. He grabs the spring onions and starts chopping them, in case Milly finishes her job too quickly.

*  *  *

"This is the best map I have of the area. You should bring it along." Father Len unfolds the map and places it on the table. His hands are shaking slightly, but his voice is under control. "Do you have a bicycle?"

"No," Ben says. "Sorry," he adds, for some reason. Tilly does not have one and he has no desire to ask Peter or Harold.

"Use mine," Father Len says. "It's in the shed. Do you know where?"

Ben nods. He has borrowed the bicycle before.

"I pumped it and oiled the chain last week, but the gears can be…" Father Len stops and frowns. Taking a deep breath, he turns his attention to the map. He points and describes the cycling route, checking with Ben that every instruction is clear. He shows him where to leave the bicycle and describes how to find the way, on foot, for the last bit. "It's easy to get lost out there." He looks at Ben with a worried expression. "Are you sure you know how to read this kind of map?"

"You taught me, remember? Last year."

Father Len nods a few times, then looks at the table again. He starts folding up the map, very slowly. He hands it to Ben, but does not let go. "Has she asked for…?"

Ben waits but nothing more is forthcoming. "Tilly told me she needs herbs for the pain and that Margot would know which kind," he says. "I think it has gotten a lot worse the last few days. She is-"

"Yes, yes, of course," Father Len exclaims hurriedly and seemingly relieved. He lets go of the map in his hand. "Poor Tilly. She doesn't deserve this." He looks at Ben. "It doesn't seem fair, does it?" He frowns again. "Did she tell you how it started?"

"She told me a bit…" Ben says, looking at the floor. "A lump. There was an operation. They thought they had gotten the whole thing."

"But you never know."

"I guess. She doesn't want to see the doctor again. I asked several times. She says there's nothing he can do."

"He can't do much. She knows that."

"But can't he give her something proper for the pain? Instead of this." Ben waves the map around. "Herbal tea won't be enough."

"These days, there's a shortage of all kinds of medication. Not all requests get approved, let alone in time." Father Len shakes his head, slowly. "Tilly is being sensible, asking Margot for help. Margot knows what she is doing; her herbs are effective. She also knows Tilly, so she will do her very best."

"So you know the... you know Margot?"

"I do. I've been out to visit her several times. We've had some very interesting conversations." He adds a brief, incomplete smile. "She's older than anyone else around here, so she has seen a lot. She's also intelligent—wise, you might call it—and inquisitive. I'm sure she'll ask you lots of questions. Be prepared."

"I'll go tomorrow morning, as early as I can."

"Good man. Will you tell her goodbye from me? Tell her I've enjoyed our discussions over the years."

"Goodbye?"

"I've been reassigned."

"No," Ben says. It comes out like a whine. He can't help it. "You can't go. You can't... I can't..."

"I'll be leaving in a few weeks. They haven't told me where to yet. A low-level administrative job, most likely—or perhaps the sanctuary."

"But why?"

"The hardliners have taken control and I'm not doing things the way they like." He sighs. "There have also been complaints about me, locally. I'm too relaxed. I don't keep to the book." He sighs again. "We have too much music."

"Too much? But who would say that? Who would complain? Is it Harold? Or Daisy?" Ben asks, fighting tears. "She is so-"

"It doesn't matter, Ben." Father Len puts a hand on his shoulder. "The papers have come through. I have to go." He draws a deep breath. "You will get someone new. But it will be someone who toes the line and reports diligently on any hint of subversive activity. So be careful, Ben." He pauses. "Are you listening?"

"Yes." Ben pouts for a while. "Why have things changed? Why now?"

"More unrest at the border, according to the reports. Some of our people have been attacked and abducted by the impure. Whole villages. Women raped..."

"That can't be true. Why would...?"

"I don't know what the real story is, Ben. This is what we've been told." He shakes his head. "Even the past has changed. On KnowledgeNet."

"The past?"

"Everyone who has ever left New Eden has now officially been abducted. They have the names and the numbers to prove it."

"Like my parents?"

"Your mother, at least. I saw the list for this parish."

"But she wasn't-"

"Don't fight it, Ben. Ignore how other people choose to see your former life. You know the truth. And don't worry too much. You are safe, I think. You are proof that it is possible to escape the control of the impure and come back home. They might come and ask you… Well, just be careful, won't you? Listen to the new priest."

"I don't *want* a new priest. I'm not going to…"

"Be there for Tilly, won't you? And for Milly. They need you."

"I know." Ben hangs his head. They fall silent. "I'll go tomorrow morning," Ben finally repeats.

They start walking toward the church gate. The sun is just down, the light soft and gentle. The flowers along the path are unreasonably pretty.

"What is the sanctuary?" Ben asks, as they shake hands. "You mentioned you might be posted there."

"It is a place for those who have only suffering left. Maybe I can finally do some good." Father Len smiles, but unconvincingly. "Or at least no harm." Ben opens his mouth to speak, but Father Len stops him with gesture and a shake of his head. "Forget that. Forget what I just said," he says and takes a step backward. "I will miss you, Ben." He looks solemn. "Tell Tilly I will come by to see her in the morning. I can help out for a few hours."

"That's OK, we-"

"I prefer to keep busy. And I *do* know my way around a kitchen."

They both look toward the small house that has been Father Len's solitary home for many years. "I suppose you do," Ben says. "I will tell her."

\* \* \*

Ben looks at the map for the tenth or twentieth time, trying to make sense of the contour lines in a large area of uniform green. This area also contains the pencil mark made by Father Len. The sun is in hiding and there are no paths to follow. He turns the map one way and then the other, trying to reassure himself. The panic only gets worse. He retraces the road he came in on and his mind returns to the bicycle. He had not asked Father Len for a lock. It seemed wrong, somehow, to ask; but it was stupid, cowardly, not to. Now he worries. The bicycle is hidden behind some dense bushes, away from the road. But what if someone were to come from the other side? It would look like the bicycle had been abandoned. What if they took it? Walking all the way back would take forever. Tilly would worry. She would be in pain. "Stop that!" he says, possibly out loud, "no one will take the bicycle". But how will he ever find it again if he cannot even find the way to Margot's house? What if he comes back late *and* empty-handed? Calm down, he tells himself, and think. The bag—it was packed by Tilly, who insisted she was feeling much better at four thirty in the morning. He checks: water, bread, even cake, a thin poncho, a notebook, the letter for Margot and yes—here, in the outer pocket—a compass. Of course she thought of that. He takes it out, lets it stabilize, aligns the map and makes a plan. Putting map and compass in his shirt pocket, he walks briskly onwards.

He arrives at the clearing an hour later and sees the old stone house with its extra-large chimney to one side. He moves closer. It is exactly as Father Len described it, including, to the left of the house, the cluster of impressively large trees. They are pine-like but not pines. The trunks are thick, thicker than his reach, he guesses, and tall, but not quite straight. Majestic, he thinks, as he lets his gaze be pulled upward.

"Wonderful, aren't they? They are my oldest friends."

He turns around to find an old woman observing him from the door. He moves toward her with a relieved smile on his face. She is almost his height, although slightly hunched. She must have been taller once. "They are," he says. He stops at an unchallenging distance. "Ben Hatton... I mean Hill." He holds out his hand and moves a step closer. "Aunt Tilly sent me. And Father Len," he adds, to be safe.

"Margot," she says, giving him a large-knuckled and liver-spotted hand to shake. The grip is unexpectedly firm. Father Len guessed her to be eighty or ninety years old, which looks about right to Ben. Her face is a busy landscape of wrinkles, slowly rearranging themselves around a welcoming smile. The thin, white hair is tied back in some sort of bun. Her eyes appear pale blue or gray. She regards him steadily. "I'll just stick with Ben, then," she says, "since there seems to be some confusion about the family name." Ben starts to explain, but she waves it off and turns around in the doorway. "Come in. I'll make tea."

Ben passes on greetings from Father Len while Margot boils water on a stove that, despite the warm weather, is putting out quite some heat. The kitchen is smaller than the one at the farm but even more crammed with stuff. Two of the walls are covered with shelves, all bursting with jars and flasks and pots and bowls. Ben explains what he can about Father Len's reassignment, which is not much.

"What a shame," Margot says. "I have enjoyed conversing with young Father Len." They move into a cozy room that appears to make up the rest of the downstairs. "And how is sweet Tilly?" she says as they sit. "Your aunt, you say?"

"Yes. My mother was Tilly's sister. Eleanor. Did you know her?"

"I did, indeed." She starts to smile, but cuts it short. "You said was. Is she no longer with us?"

"No." Ben puts his cup down. "She died. Four years ago. In an accident."

"I'm sorry to hear that." Margot looks like she means it. "Tilly and Ella used to visit me with their mother. The girls were close in age, but so different in temperament. Tilly was careful and watchful. Ella was sassy, always full of spark. She was also very pretty. But I guess you know that." She looks at him with a kind smile. "She crossed over, didn't she? Joined the heathens."

Ben nods. "Yes, she did. I grew up there."

"How exciting!" She rubs her knobby hands together and adds a small shudder. "I want to hear all about you and your life over there." Margot's smile becomes mischievous. "I'm afraid most people around here consider *me* an irredeemable heathen. I suppose you have heard that."

"Some of the people in the village call you a witch," Ben says, trying to sound nonchalant. "But I trust Father Len and Aunt Tilly."

"Good, good." She chuckles. "But first you must tell me why you are here. In case I need to get something going for you to take back."

"I…" He stalls and looks at his cup. "It's Tilly." He looks at Margot again. "The cancer has come back. She is in pain."

"Oh, no. Poor Tilly. Sweet Tilly." She shakes her head slowly. After a moment's pause, she speaks again. "So she needs…"

"Something for the pain. She said you would know what."

Margot nods and waits. When he does not add anything, she puts her cup down and thinks for a while longer. Then she gets up slowly. "Let's go back to the kitchen," she says. "I'll give you the dry mixtures, but I also want to boil up some extract for a flask, ready to use. Shall we?" She leads the way. Ben follows.

For the next half hour, Margot works away busily, explaining as she goes along. Ben gets the notebook and a pen from his bag and writes down her instructions. She weighs, chops and mixes. She throws some of it into a small pot on the stove. The rest is portioned into drawstring satchels. She labels them 1, 2 or 3. Finally, Margot wipes her hands, lifts the lid of the little pot with the cloth of her skirt and gives the mixture a stir. "This will be ready in an hour, two is even better." She points to the satchels. "Take these. It should be enough for three to six weeks, depending. I'll go collecting tomorrow so I can make more." She frowns. "Come back for it—or whatever else Tilly needs—any time. I'll be here. Poor Tilly," she repeats and sighs. She turns around and moves back into the living room. Her movements are slow and careful. Ben puts the satchels into his bag and finally remembers the letter from Tilly and the carefully wrapped pieces of oat and honey cake. He brings both to the table and puts them down in front of Margot.

"Let me refresh the tea," he says and reaches for the teapot. Margot makes no polite protest. She simply nods and picks up the letter.

Ben takes his time making the tea. When he comes back, Margot has finished reading the letter. A couple of loose sheets filled with Tilly's fast-flowing script lie half folded on top of the envelope. Margot's gaze is distant. He refills their cups and puts down the two small plates he found for the cake.

"Thank you, Ben." She smiles at him. "Tilly tells me how happy she is that she has gotten to know you—and that you have been a great help—still are—I can see that." He smiles back. Once seated, he reaches for his cup. She puts her old hand on top of his. "You have done some good here, Ben. Remember that."

He nods and blinks rapidly.

"So." Margot's face lights up. "Tell me everything about life on the other side! I am *so* curious. I've heard a few things about the chartered way, but… tell me." The switch is so abrupt that it takes Ben a moment to find his voice.

"Sure…" He hesitates. "But do you think… Is there anything you can tell me about my mother—about Ella? I'm trying to… understand."

"I'd be happy to, but I don't know much. They only came to see me a few times. And when I last saw her, she was still a girl—well, a young woman."

"How about Aunt Vera?" Ben suddenly remembers. "I mean Vera Weiss. She wasn't actually my aunt, but… Tilly told me you knew her."

"Vera Weiss?"

"Originally Vera something else, according to Tilly. She was somewhat older than my mother, I believe, but also came from here."

"Yes, of course, Vera…" Margot smiles, but hesitantly. "My proud and brilliant Vera. Vera Birch, back then. I suppose I knew that she had changed her last name, but…" She pauses. "Is she…?"

"She died—a couple of years before my parents."

"Oh…" Margot seems to drift away for a moment. "I suppose I shouldn't be surprised. I'm such an old crone… too old…" She shakes her head, with sadness or weariness or both. "So, yes, I did know Vera. I knew her quite well." She smiles faintly. "She lived with me for a while when she was a teenager. Can you imagine that? All the way out here? But she didn't complain. Later, when she was studying, she'd come back to visit me. That was when Ella and Tilly met her. Ella was in awe of Vera; I could see that. She idolized her."

"Vera studied here?"

"At the University. She studied medicine."

"So that's possible? I thought my mother couldn't…"

"Sure it is. We may not have as many fancy techniques as in your world, but doctors are always needed. Your grandfather didn't approve of women doctors, I think, and that's why Ella couldn't go. She must have tried to convince him, I imagine, but… It was different for Vera. Her father died before she finished school. Her mother had long since passed, so she had to move… Let's just say it was complicated. She ended up in my care for a while. She helped me out—in the kitchen." Margot winks. "She was a natural."

"But your-" Ben frowns and gesticulates toward the kitchen, "type of medicine is not that of a medical school."

"Vera learned both. Whatever works, is what I say." She sighs. "Like Tilly's operation. That worked—for while. But sometimes—often—the hospitals don't have the drugs they need." She gestures toward his bag, standing next to his chair. "And sometimes they can't do what is needed."

"Can't do?"

"Well, like everything else, they are controlled by the church. They can only do what is sanctioned. I'm… I help people with whatever they need."

Ben looks at Margot, carefully. "Including things the church is not too keen on?"

"Yes." She nods, slowly.

Absentmindedly, he eats some cake.

"Tilly was shocked when she heard where my mother worked," he says.

"Which was?"

"A hospice. Where people go to die. When they decide to die."

"So that *is* true? That people choose their time to die? I have heard this before, but I wasn't sure…"

"It is." He pauses. "Vera did it, as well." He pauses again, his eyes resting on Margot's face. "Cancer, you see." Margot nods, calmly. "Even with the protective edits and the treatments and everything, it still… and I guess it's very…" He frowns. "Most people get quite old. But eventually, they feel that… they find peace with…" Ben realizes he is slipping into a defensive mode. He tries for something more factual. "It's always your own choice. Children cannot decide for their parents, for example. And the procedure is never done at the spur of the moment. My mother told me-"

"Ben, you don't have to convince me," Margot says, gently. "I approve. I think the most important choices should be your own to make. About your life *and* giving life. No one should be forced to suffer. No one should be forced to have children they don't want. It seems straightforward to me." She looks Ben in the eyes, steadily. "Maybe now you understand why they call me the witch."

He hesitates. "But they let you be?"

"They tolerate me—way out here—until they don't. I have been run off a few times. Not burnt at the stake, thank God." She smiles, another mischievous smile. "We are not barbarians, you know."

The comment triggers the beginning of a memory for Ben. He tries to get a firmer grip on it, but it slips away, leaving only a vague sense of loss. He shakes it off. He considers what else to tell Margot about the society he grew up in. Should he describe the net and its distractions? But how can he explain all that? And would Margot be interested? "You said you'd heard some things about life over there," he says, instead. "What did you hear?"

"Well… Vera managed to get a few letters to me, early on. They were sent with medical shipments, I think, through some old friends of hers from University—some complicated route. She wrote to me about what she saw and learned. I particularly liked the voting system. From age twenty to eighty, is it?"

"That's true. Working life, essentially."

"I love that!" Margot exclaims with unexpected glee. "I'd be past voting age. More to the point, so would many of those awful old priest who hang around forever and control everything." She chuckles. "Be gracious and step aside, please. You have had your time. Ah, wouldn't that be something?" She shakes her head. "So, when do people usually have children?"

"In their thirties and forties, normally." Ben remembers the children from the village and their many questions. "But they are not from incubators. Mothers still carry babies—the old-fashioned way. It's still the best way, it seems."

"Well, evolution has had millions of years to get those things working optimally." She notices his expression. "Oh, don't look so surprised," she huffs. "We aren't all ignorant. I also studied medicine for a while. I was the one who encouraged Vera—and your mother, later on. What I do here is a natural extension of that, I think." She pauses. "But children aren't conceived the old-fashioned way, are they?" She cocks her head,

"Sometimes they are." Ben feels himself blushing, stupidly. Margot sees this and smiles, indulgently. "But most are not," he continues. "In vitro fertilization is more popular. In fact, IVF used to be obligatory, to make sure everyone got the HQV resistance alleles—or edits."

"So, tell me: editing, selecting, cloning—or youngtwins, is it? What do people really think of all that? We mostly hear the horror stories from the time of the crisis."

"Human cloning. That was from adult cells. It didn't work very well." Even the version of history he was taught acknowledges that.

"I know it's different now. I understand what youngtwins are. They're like identical twins, just not born at the same time. I'm not sure it's a good idea but…"

"It's an abomination, is what I've heard." He says, flippantly, and then immediately regrets it.

"Not from me, you haven't," she snaps.

"I'm sorry," he responds quickly.

She nods. "So, I want to understand this editing business," she continues, in a normal tone. "Vera didn't explain properly."

"I'm no expert."

"But you know the basics."

"Of course—from school. And a couple of my friends were really into biology." Memories of Eiko and Raphael flash by. They are explaining, discussing and laughing. It is so unexpected, and so life-like, that he is momentarily stunned.

"Explain it to an old fool," Margot says.

Ben explains, as best he can. He starts by talking about the HQV resistance edits designed to prevent virus infection. He describes how they were first tested: the original editees, the famous five. Margot is curious about them and also about their parents, who made such unusual decisions. This allows Ben to tell the story of Aurora, one of the five and his mother's most famous patient. He goes on to explain the voluntary edits that are now possible—disease related or not—and the profiles used to favor more complicated traits. He also mentions the blood stem cells and somatic editing used to treat cancers and some other diseases, but acknowledges that he does not know the details. At this point, Margot has stopped asking questions. Ben thinks she may be tired and stops talking.

"It must be difficult," Margot finally says, looking pensive. "I realize that. So difficult."

"What?"

"All that choice. What do you do with it?" She gestures. "Avoiding disease genes is straightforward, I suppose. But apart from that… Do you want a boy or a girl, a tall one or a short one, blue eyes or brown eyes, strong or lean? A math genius or a genius on the violin?"

"Genius is a little tricky. Even talent is hard to pin down."

"I guess it must be."

"And not everyone takes all the options. You don't have to."

"But still—you can't un-choose choice. It'll always be there, won't it, between parents and children? What they chose—and didn't choose."

"Yes," Ben answers. He thinks of his friends again and feels a strong urge to tell Margot about them. But something stops him. Instead, they talk about Vera: How happy Margot had been to have this bright and lively girl join her household of one, her shocking decision to leave, the tough transition and her well-respected contributions in the new world. Ben is grateful that Eiko got him up to date on those things. He also describes her role as Aunt Vera, his parents' closest friend.

Margot finally puts her hand on his and closes her eyes for a moment. "Thank you for all that, Ben," she says. "To know that my Vera has left her mark, a valuable mark, on the world and on your family, that makes me content."

Ben can see that Margot is getting truly tired. When he asks, lightly, about the extract she is boiling up for Tilly, she pulls herself together with some effort. They move to the kitchen, where Margot takes the tiny pot off the stove and sets it to cool on a stone. She directs Ben to collect a ceramic flask from a high shelf and takes out the stopper. Then they stand, side by side, and look out through the open kitchen door. The cloud cover is only partial now, but the forest looks dense and dark beyond the clearing. He knows he should be going soon. Before he does, he decides to ask the one question that has been bothering him, in one form or another, for so long.

"I probably shouldn't be asking you this, but… Aunt Vera took the legal route: She had somatic edits and qualified as a chartered doctor. Why didn't my mother do the same? She was still young when she crossed over. She had Vera as an example and a friend. Yet she stayed in the shadows, as an illegal. She became bitter. Why? Why so ambitious for me, but not for herself? It doesn't fit at all with the feisty Ella you and Tilly have described to me."

"Your mother was more conflicted than Vera, I suppose. Some of this you must have worked out by now. How old are you?"

"Twenty-two."

"And when did Ella and Jacob cross over?"

"I'm not sure. Thirty, thirty-five years ago?"

Margot looks at him and frowns. "How do you get that?"

"Tilly said she was about my age when they left. I figured she was mid-fifties when she died. She never said, exactly, but the other parents…"

"She left New Eden twenty-two-and-a-half years ago."

"No!"

"Yes, Ben. I saw her just before she left. Tilly was with her. They had come for one of my potions. You can guess which one."

"No…" Ben is still shaking his head.

"Apparently, she couldn't go through with it."

"But…" Ben goes quiet. He finds nothing more to ask. He should have worked this out much earlier. He has no idea why he didn't. The clues were all there: His parents' vagueness about their age, Tilly's odd fudge about his age when he first arrived and—most obvious of all—his own damn DNA sequence, the secret kept so well by Aunt Vera. Now it all made sense. His strong-willed and stubborn mother, she did all this just to-

"The only thing I don't understand," Margot interrupts his thoughts, "is why they left Tilly behind. It took the life right out of her."

"What?" Ben looks horrified. "What do you mean—left her behind? Tilly told me she never wanted to…"

"That's our Tilly, isn't it?" Margot gives him a sad half-smile. "She is the kindest person I have ever met. Of course she didn't tell you that part of the story." She sighs. "Tilly picked herself up, after a while. She made the best of it." Margot feels the side of the cooling pot and places a tiny strainer over the flask opening. "Hold the flask, will you?"

Ben does as he is told. Margot pours expertly and pushes in the stopper. Ben picks up the flask. It is quite warm.

"Take care of her, won't you?" Margot says. Ben nods. "Give her my love. And come back when you need me again."

*   *   *

Tilly grips his hand. Her face contorts as she squeezes it hard.

"Do you need the flask?" Ben asks, suppressing a wince of his own. "There's still some left. I've also started brewing the next batch." But it is medium strength, he remembers. They have used up the number three satchels. He will have to go back to Margot as soon as he can. Tomorrow morning, if Tilly can spare him. His anger at Harold and the rest of the family flares up again. They never sit with her. They pretend not to hear her screams. Daisy and Anne are occasionally seen in the kitchen now; that's the only change. He'd prefer it if they stayed away altogether, but it does give him more time for Tilly. She seems to be saying something. He leans in closer and sees the tears running down her cheeks. The whisper is too faint for him to hear.

Ben feels a tug at his free hand. Surprised, he pulls it back quickly. He turns his head to see Milly standing next to him.

"Sorry, Milly," he says. "I didn't hear you come in. Did we wake you?" He put her to bed hours ago and remembered to close the door once she had fallen asleep. But she sleeps lightly, now. What does she understand, he wonders? Does she know everything will change soon? He looks at Milly's face. She is looking at Tilly, whose eyes are closed. Milly's face does not show terror, only concern. She hands him the ceramic flask.

"Thank you, Milly," he says, placing it on the bedside table. "And a spoon, could you get me a spoon?" She runs out and returns a moment later with a spoon, which she hands to Ben, and her favorite stool, which she places next to Tilly's bed.

Why not? Ben thinks. Who am I to say what is best for her? A faint smile settles on Tilly's lips when she feels Milly's small hand on her forearm. She must have heard the exchange, because she lets go of Ben's hand and opens her mouth a little. Ben spoons in the last of the extract, making sure not to spill anything. He places his hand back where Tilly can grip it, which she does, but less tightly than before. The three of them sit like this for a while.

"Ben," Tilly says, "you must go to see Margot again." Her voice sounds surprisingly normal. Her eyes are open and her face is calm.

Did I fall asleep? Ben wonders. Milly is still on her stool, fully awake, apparently having a silent conversation with Felix, a toy dog with very long ears.

"More of the strong number three and…" she takes Ben's hand, "ask for the mercy potion as well."

"No," Ben says quickly, "not… not yet."

"Yes, Ben. Please." She squeezes his hand. "We spoke about this. It is what I want." She closes her eyes for a moment. They *did* talk about it, after a long night where the pain would not stop. This was before they switched to mixture number three. He remembers being surprised at her request. Right now, he does not remember why. "Pain is not ennobling, Ben," she continues. "It is horrible. I have had enough."

"I will go in the morning," Ben finally says. "With the bicycle, I can make it back before dark. She should have more of everything now. She knows…"

"She does. I told her in the letter." Tilly pauses. "But don't tell Father Len what I have asked for. It is… difficult for him."

"I won't."

"I hope they don't send him to the sanctuary. He will go if they tell him to. But it will break him." Ben is still not completely sure what the sanctuary is, but he can guess. He knows that Tilly refuses to go to there.

They are silent for a while. Only Milly is moving a bit, animating her dog.

"I was so in love," Tilly says, out of the blue. "So much in love. It was silly and it was hopeless. But it was real."

"In love?"

"Yes." She smiles. "I couldn't help it."

Ben looks puzzled. "Who?"

"Jacob, of course."

"My father?"

"You should have seen him—those dark, dark eyes. And to dance with him, it was…" She sighs. "He was like no one else. Fun. And gentle, as well, like you. But Ella was too beautiful. I had to let him go." She smiles again, nonetheless.

Ben sighs. Maybe. Maybe that was how it was. He never got around to asking Tilly about Margot's version of the past. The hurt they caused. It seemed—so long ago—and so… not for him to know, really.

"Have you ever been in love, Ben?" Tilly says quietly, looking straight into his eyes.

"I don't know. I…" He looks down. He waits a moment. Then he looks back at Tilly's face. "Yes. I think so." He smiles. She smiles.

"And?"

"She was my best friend's girlfriend." He shrugs. The memory is both nearby and impossibly far away. That last evening, under the stars. He does not remember exactly what he said to her. "It wasn't to be," he says to Tilly now. She nods.

"Tell me a story, Ben," she says after a while. Milly looks up. She never misses a word, Ben realizes. "Tell me about Jack and the maverick bot." She slides further down on the pillow, her eyes half-closing. "Or about the doll-maker and his beloved. Yes. The doll-maker."

\*   \*   \*

The sky is lightening. Ben looks at the low line of clouds, a glowing, vibrant red below gentler bands of orange and pink. The silent display, shifting ever so slowly, is beautiful. For the moment, he tries to not to think about anything else. He simply walks along the empty road and looks to the east.

A short while later, he is surprised and a bit worried when he sees a light on in one of the windows. I cannot lie to him, he thinks.

He looks though the window. Father Len is sitting at his desk, his back to Ben, and seems to be writing something by hand. The last sermon, Ben guesses. The last one in this church, anyway. He sees the boxes and the half-empty bookcase. He looks away. Heeding Tilly's request is suddenly the easier option. He goes straight to the shed and grabs the bicycle. It is unlocked, as always.

There is enough light to navigate along the old road. He grips the handle-bars tightly and pedals hard. The flow of air dries his face. The physical effort calms him. Concentrating to avoid potholes and wind-blown branches keeps his mind busy. When he reaches the bend in the woods, the sun has already been up for quite a while. He finally realizes how thirsty he is.

He takes the map out, and the compass too, but finds that he does not need it much. As he walks, he starts rehearsing what he will say to Margot. He real-izes that he never asked her about her parents, or about the early years in New Eden. She must have seen so much. He tries to focus on that time, imagining what it must have been like and coming up with questions where he falls short. It works quite well. He is almost sufficiently distracted.

Something seems to be different in the woods. As he gets closer to the clear-ing and to Margot's house, he realizes what it is. There is a path now, a line in the undergrowth trodden flat by passing feet. He starts walking more slowly, taking care not to make noise. He feels a tightening in his chest. Fear. When he sees the house, he stops completely. It looks the same as before, he thinks. Next to the house, he sees the grand old evergreens, unchanged. He looks back at the house. Not the same. The door is ajar.

"Margot," he yells as pushes the door fully open, "it's Ben. I've come for-"

He stops and stares.

The kitchen has been ransacked. Broken jars are everywhere. Their contents cover the floor. There may not be a single intact vessel left in the room, save the iron pots. The cover is off the stove. As he sees this, he also realizes that the place smells foul. He steps over to the stove and peers down the hole. It is full of ashes and half-burnt plant parts. Although the stove is cold, the smell seems to be coming from there. He thinks this is probably a good sign.

"Margot! Margot, are you here?" He yells as he walks through the room they had their long, meandering conversation in—just a few weeks ago. He yells again as he peers up the back staircase. The living room looks almost normal, with only a few things scattered about, mainly the contents of a small bookcase. Moving about slowly, he tries to imagine what happened here, pre-cisely. He stands for a while in the middle of the room as he considers the need—or not—to go upstairs. It takes him a while to decide.

In the end, Ben does go upstairs. But no one is there. He does not know if it looks any different from before.

Later on, he will find that had he spoken to Father Len in the early hours, he would have saved himself the trip. The news had come to the priest the day before. In the morning, at his writing desk, he had been struggling with how to transform his outrage and his sadness into a sermon they would remember.

Later on is also when he feels Tilly clutch his hand more desperately than ever. He finds himself wishing she would crush it, squeeze all pain out of it. But Tilly does not have the strength. He should have gone to Margot's place earlier; he should have asked for it the first time; he should have… Tilly tries to tell him that it is not his fault. But Ben feels the truth.

# 8

## Business Acumen

"It would be more convincing without the visual. His tell is too obvious."

Victor gives a small jolt of surprise. The voice is very close to his left ear. Then he remembers that an attractive young woman had placed herself to his left when the meeting started. So, he responds, in a whisper. "His tell?"

"When he lies. Like if he's asked about a decision they've already made but has to pretend they haven't. His mouth twitches." She draws back a bit. "Right side. Watch."

Victor watches as the discussion continues. The topic is dome security. The focus drifts somewhat. A representative of the new Peaceful Union Party asks whether the controversial track and freeze approach will be implemented. The minister answers that this has not yet been decided and that he appreciates the complexity of the issue. The corner of his mouth twitches. Victor wishes they could get the PUP people out of the room, so the rest of them could get on with things. Then he remembers the whispered comment. "Well spotted," he says and looks at the woman to his left. She has a striking face: large green eyes, high cheekbones, flawless skin and generous lips, now curled into the faintest of smiles. She is also very young. He wonders how she got invited to such a high-level meeting.

They find each other at the coffee, where the twenty or so in-person participants are doing what the video-linkers cannot: old-fashioned networking and skillful scoping of territory.

"Victor Huang," he says holding out his hand. "Of Huang Shields."

She smiles at the superfluous addition. "Celia Saunders." She pauses for a moment with a hint of teasing. "Of Response Modes."

"Response Modes?"

© Springer Nature Switzerland AG 2020
P. Rørth, *The Unedited*, Science and Fiction, https://doi.org/10.1007/978-3-030-34624-9_8

"It's a new company. We do response predictions, primarily for large groups, human only or human and AI mixed."

"Interesting."

"The government seems to find it useful."

"For dome security?"

"Security has many levels. Most involve people."

"True. I've just never heard of behavioral predictions being used in this particular area." He smiles. "And I tend to hear what's going on."

"We are the first." She pauses. "With our particular approach."

"And your role is?"

"Product development and marketing. It's a very small company."

"But you are here today."

"We've got a couple of government contracts already. Minor ones. The public security business is new for us. We're just dipping our toes in."

"Economics and the financial sector are more familiar arenas, I presume. Do you think trading logic transfers to conflict and crisis predictions? Escalation assessments? Or is this contingency planning?"

"Everything is contingency planning these days, is it not? But yes, our foundation is in economics."

"I'd love to hear more." He smiles, a relaxed and confident smile.

He is about sixty-five, Celia guesses, maybe seventy. Lean but not thin, the hair defiantly gray but cut very short.

"Would you care to join me for lunch afterwards?" he asks. A small twitch pulls at his mouth. "Or rather, join us? I'm having lunch with my son at the penthouse restaurant." Another twitch. "He's supposed to update me on some business details. But that can wait."

Celia's initial response, excitement, is quickly tempered by slight irritation and finally rendered smoothly professional. "Of course," she says. "It would be my pleasure." Victor grins. Celia guesses he has interpreted her layered reaction as she intended.

In the next session, both wall and dome security are discussed and Victor Huang gives a short presentation. He is not introduced and remains seated for it. His description of Huang Shields' approaches is heavy on confidence and short on details. Despite this, Celia senses that his authority is well earned, based on up-to-date knowledge as well as long experience. A few key numbers and technical terms sneak in and the superficial platitudes are few. She asks a question about a hypothetical scenario she has been working on. He thinks for a moment before answering thoughtfully, but succinctly. He remembers her name. Later in the session, she finds occasions to pose insightful and pertinent questions to other presenters. She guesses that this increases Victor's interest in her rather than turning it off. She is right.

*   *   *

"Hello, father," says a voice approaching the table from behind Celia. "I thought we were going to …" When he sees Celia's face, he stops abruptly. Victor misinterprets the cause of his son's surprise and starts the introductions with an only partially suppressed smirk. "Leo, this is Celia Saunders. Celia, my son, Leo."

"Oh, we've met before," Celia says before Leo has a chance to speak. His eyes go wide in alarm. "A couple of years ago. It turns out we have friends in common." She smiles at Leo, a notch too sweetly. "At a dance club, I think it was." She catches his eyes and does not let go. "Very nice to see you again, Leo."

"Yes, of course." He swallows. "And you, Celia." He waits for another moment before taking his seat.

"This is delicious," Celia says, turning her attention to her plate. "You should have some. Your father is paying, I assume."

"Naturally." Victor says, with satisfaction.

They end up having a lively lunch, discussing the analyses pioneered by Response Modes as well as shields and security. Both sides reveal enough to keep it interesting, while maintaining a cautious distance to sensitive details. Celia explains how the models for virtual and physical crowd responses were developed from models of trading behavior in the financial sector, but leaves out the programming shortcuts that would key Leo in. The Huangs talk about new shield types produced and others being developed, but only in terms of intent and success rates, not exact properties. Leo soon forgets the initial awkwardness and regains his normal voice, eager and competitive. He works on software development for some of the most advanced shields.

While they talk, Celia looks back and forth between the elder and the younger Huang, comparing their faces, bodies and gestures. The similarities are partially obscured by the age difference as well as by hair color and physique. Leo obviously still works out. But somehow, the likeness is all the more striking for this tension, and, like an unexpected revelation, it sticks in the mind.

At one point, Victor excuses himself. Leo immediately leans over toward Celia.

"How come you didn't tell him?"

"Tell him what?" she answers with mock innocence.

"About the trip. About my many ill-considered remarks at the time."

"Why should I?" She cocks her head. "We were all blowing off steam back then, one way or the other. We were young."

"Were? It was only two years ago."

"Some of us grow up fast."

"Speaking of … I thought for sure you'd be at university."

"I was. I am. This is just a hobby." The deliberately blasé comment seems to get to him, she notices. "No, just kidding," she adds quickly. "This *is* my real job. I'm actually … Well, whatever. I do uni a few mornings a week. I'll get the degree sometime. Or not." She shrugs. "And how about you? I have to admit this is the last place I expected to find you: with your father—and working at Huang Shields."

"Well, things change. As you said, we grow up."

She looks at him for a while, challenging him with an expression of overt skepticism. He looks right back, neutrally, not giving in.

"The hack of the Reserve database," she says, abruptly, "did anyone ever find out that was you?"

"No," he hisses, glancing around quickly. The tables are far enough apart for privacy, he judges. He continues in a half-whisper. "Only a few people know. You guys—and our hosts at the transit station."

"And they never leaked it to the relevant authorities?"

"Apparently not."

"You were lucky."

"I was smart." He grins. "I covered my tracks well. I also set it up so that only a small nudge was needed for the hack to trace straight back to New Eden. Being a nice guy, I promised not to nudge anyone unless provoked."

"So they kept quiet. And the rest of us were too …" She realizes one person might have wanted to tell. She does not want to go there. "Well, other things quickly became more important, didn't they?"

"Yep," he says with a big smile.

His apparent pleasure in the situations makes no sense to her—until she remembers why it was that he was so angry back then. She shakes her head at the incredulity of it. "You are quite something," she says out loud.

"That's me." He grins even wider, ignoring or not catching the ambiguity in her remark. He sees his father returning across the room. "Let's meet up sometime," he says, quickly. "I know a really cool place." He raises an eyebrow. "It'll be fun. I promise."

She stifles a laugh. "OK," she says. They exchange contact details with quick taps on their wrist-links.

Shortly after this, Leo is sent back to work. Victor and Celia descend thirty floors for the continuation of the meeting. Before they join the group, Victor touches her arm lightly to get her attention.

"You know that Leo is my…" he starts.

"Youngtwin? Oh, yes. I've known that all along." She shrugs. "As far as I can tell, this means that he is much like you, but without the experience or the authority." She looks at Victor's face steadily. "Meeting him reminds me that when it comes to men, the twenties are by far the least interesting decade." She gestures with an expression of exaggerated regret. "Too much empty pride and too much crude lust." She notices his immediate reaction, half shock and half fascination, before it is concealed behind an amused smile. "The question is only how well they hide it," she continues calmly. Before he can respond, she continues. "I'll give you my contact info." She holds up her wrist-link. He does the same, more or less automatically. "Just in case you get seriously interested in Response Modes," she adds and taps the link quickly, expertly. And then she is gone, across the room.

\* \* \*

The place is hopping. It has been, for hours. When they first came in, she got the usual appreciating looks. It still pleases her, but only for a moment. It is too predictable, too boring. The massive dance floor is, however, not the least bit boring. The beat of the music is powerful and persuasive; her body responds instinctively. Some of the tunes have words, banal but catchy; she sings along happily. Yes, she thinks, this was definitely worth going out for. Then she remembers Leo. She looks around. The many dancing bodies close by do not restrict her movements directly, but do provide a pleasurable crowd effect. She could model this effect, she thinks, and optimize it. She chastises herself. Never quite let go, do you? The abandon of youth, but with a lower-case a. Not a bad place to be, actually. She touches the little pocket in her non-transparent top: a useful hiding place for anything she is unsure about. Currently, it contains three mystery pills. To her, this is a practical and innocuous solution. She prefers control. She always has. Leo, on the other hand, does not look like he has a pocket for caution. He is moving closer, dancing the whole way. She admires his style: energetic, sexy and joyful. She falls in with him. The music glides into another tune, and another.

"I'm done for," she finally acknowledges. They move off the dance floor. She grabs two bottles of water and hands him one.

"But you had fun?" Leo's voice is too loud now that they are away from the music.

"Absolutely." She gives him a big, and genuine, smile. Not enough dancing in her life these days, she thinks.

"I knew you would." He leans toward her, unsteadily. His eyes are glassy. "You are beautiful. Do you know that?"

"I know." Celia smiles again, but overbearingly.

She takes his arm and aims for the exit. No longer propped up by the music and its rhythm, he has become more wobbly. But he responds to her guidance.

"Hire-pod?" He asks once they are outside. There are lots of them lined up, all waiting for late night customers, all with manual drive safely disabled. She studies his flushed face for a moment. His eyes slide around, unable to keep a steady focus. His smile is quite charming, though—open and hopeful, almost innocent.

"Sure," she says. "Your place." His smile broadens.

The pod responds to Leo's wrist-link and "home" command; it starts up, quietly and smoothly. They glide past dark buildings, lonely streetlights and all-night advertisements. They do not go very far, but by the time they arrive, Leo is fast asleep. Celia is not surprised. She rouses the uniformed doorman with a well-calibrated mix of apologies and charm. He helps her get Leo out of the pod and into the lobby. She explains about the unknown quantities of party drugs and wonders out loud whether it might not be safer for her to stay and keep an eye on Leo. The doorman notes that it is rare to find a girl so pretty and yet so sensible. In the same flattering tone, he requests her ID. Without this, she cannot stay. Of course, she says, as she slides her wrist-link over the sensor on his desk. She reminds herself that there will be a log of this. The doorman helps Celia get Leo upstairs and into the apartment. He leaves with polite wishes for a good night.

Half an hour later, Celia is sitting at Leo's desk. It is at the far end of a large, open living space, facing the corner windows. Outside, the sky is still dark. Leo is on the sofa where Celia and the doorman left him. He now has a blanket for cover and is snoring loudly. If she is lucky, she may have enough time. She activates the computer and gets started. She needs his wrist-link and ID at various points. The remote pad is convenient for this and he seems oblivious to the light touch of her fingers on his arm. Whenever his snoring stops or changes, she blacks the screen and checks on him. So far, he is nowhere near the surface.

The initial layers of security are fairly sophisticated, as she had expected. Using every ounce of available concentration and imagination, she nevertheless manages to make her way through them. She is certain that he could do even better than this and that she is succeeding largely because he never expected an equal to enter his physical space. The lower level files have limited separate protection. This is a welcome gift, as the sky is now lightening considerably. She manages to sift through whole directories with simple search terms. Unexpectedly, the desk has a pad of notepaper and an actual pen. She decides to make a running list of all the wipes she will need to do when she is done. She rips the piece of paper off the pad before writing anything down.

"Well, hello there," she says when she sees the message. She glances quickly toward the sofa. Nothing. He must not have heard. The message, in fact the whole folder of text messages, is in plain text, not encrypted. The arrogance of the great hacker, she thinks. He still uses the formal title, which makes it even easier for her. It is funny, or sad, perhaps, when you think of it: Father Elias.

She had not known the actual form of the deceit, but she knew it had to be there. Love can turn to hate in a flash. But hate back to love? Or even like? Even in a lifetime, that is hard. At least, this is what she has heard. She does not have any personal experience of hate. Indifference, she knows. Contempt, she will admit to, if pressed. But not the kind of hot, focused hate and rage she saw in Leo two years ago. And now? Even at their first meeting, the casual, but expensive lunch, she could tell something was off. It was too smooth and too easy. She does not believe in easy. She knows that some people, Rafi for sure, would see an irony in that. They do not understand. Nothing of significance is easy.

She looks at the message to Father Elias again. She finds quite a few more. One has numbers in it. It looks like bank details. She writes another paper note to herself. Maybe emotion is secondary here, she thinks. The main motive could simply be greed. The next message rattles her, momentarily. Did this really happen? It hasn't been on the news. She tells herself to move on, closes all messages and starts the careful wiping. She prefers not to give Leo any reason to check his cameras, wherever they are, too soon. Just in case, she also leaves an active trail of hours spent on "Wall Street", her favorite online game.

When she is done, it is completely light outside. It is yet another gray day, by the looks of it. Leo's breathing is getting lighter. She leaves him a note and tiptoes out. She makes sure to give the doorman a generous dose of charm as she passes his desk.

\*    \*    \*

"So, you are co-founder of the company?" Victor smiles and dips his head slightly. "I am impressed."

"You looked me up?" Celia says, but smiling as well.

"Of course. I was curious."

"And what else did you find? University incompletes?"

"Self-made money for the start-up. I respect that."

"A few good guesses and a bit of luck."

"Stocks?"

She cocks her head, still smiling, but does not answer.

"So your partner in Response modes, Wang, he does the computing?"

"She, actually… Yes, she's a complete whizz. We make a good team."

"Hmm. This is very fascinating." He scrolls through the graphs, schematics and animations on his fold-out screen. "I'm trying to come up with a way to use it in our work. An excuse, really." He smiles, but does not look up. They are having lunch in a ground-level restaurant close to the offices of Huang Shields. The food is excellent, Celia thinks, but the location is uninspiring. She is reminded of the penthouse restaurant with its fantastic view. So much has happened, so quickly. First meeting the impressive Victor Huang in person and sparring successfully. Then meeting Leo again—his apparent transformation—her nighttime discoveries… She looks up. Victor is not scrolling any more. He is looking at her.

"So, Leo took you out?" Victor asks, as if he had read her mind. He looks at her face, scanning for reactions. Celia recognizes the hints of insecurity in this and finds that she likes him the better for it. He is only human. He must understand that he created an impossible situation for himself some twenty-three years ago. Why did he do it, she wonders. Arrogance? Probably. It can be as blind as stupidity. And now he has to deal with the consequences. Which is the least problematic: the younger version angry and estranged—or close at hand, a constant reminder of his own distant youth? He is the closest possible relative, yet still another, an unknown. In fact, he is a ticking bomb. But she has no intention of revealing that.

"He did. We went dancing. It was great fun," she says, in a breezy tone. "But he overdid it at the club. I had to take him back to his place and even get the doorman to help me get him up. He was out cold. Exhaustion, perhaps." She shakes her head. "Well… That was the least I could do for a good night out. Now I've had my fill for the next half year."

"Your fill?"

"Of loud music. I'm back to focusing on work. My real passion."

"It is?"

"It is," she says and looks at him steadily. "So Leo told you?"

Victor hesitates. "No. The doorman."

"He was very nice. The doorman, I mean."

"He is… loyal."

They wait for a bit, without words. She finishes her main course and sips the wine. He has already put his plate aside but joins her with the wine.

"I really appreciate this, Victor. First of all, another gorgeous meal," she says, smiling, "but also you taking the time to see me and to look at our material."

"What else is money for? And time?"

"I suppose." She takes another sip. The wine *is* very good. She could afford it herself, she knows. But she is in a different phase: The building phase. And even that… She puts the glass back down, straightens up and smiles with what appears to be some reluctance. "Actually, Victor, there was another reason why I wanted to see you. I have some questions for you." He looks wary. "There are a couple of hypothetical scenarios that we've been asked to do Response Modes on. Government stuff. They involve the use of shield technology, so I feel I would be a fool not to consult you." She ups the smile.

"And it's not a problem to…"

"Oh, no. I can talk about it. One nice thing about hypothetical scenarios is that anyone can think them up and work them up—without actual intent. You can work up half a dozen contradicting ones, if you want."

"I'm all ears, then." He still looks wary.

"The first question is about shield functionality—well, essentially about penetration." She maintains a business-like tone and refrains from looking at him too directly. "I assume shields are never absolutely perfect? That's why new ones are constantly being developed?"

"Yes. Any shield will have weaknesses. Physical or… software-related. Once they are uncovered and understood, shield-zappers and shield-spears can be designed. The former inactivate, the latter… breach the shield. That's why staying at the forefront of development is essential."

"In your presentation at the meeting you said that the newest, most sophisticated models were fundamentally different from their predecessors." He does not respond, so she continues. "This includes the larger shields, I assume, for the domes and the upgrades on sensitive city areas. In what way are they different?" Again, she waits for a response. "Or was that just a sales pitch?"

"They *are* different." He draws a deep breath. "Obviously, I'm not going to tell you exactly how."

"Obviously." She nods once, slowly.

"Suffice it to say that in previous upgrades, our in-house wrecking team," he gives the last words an ironic emphasis and smiles, briefly, "was able to develop zappers and spears very quickly."

"In-house wreckers?" She mirrors his smile.

"They keep us on our toes." He raises an eyebrow. "They are *very* good."

"And with the new models?"

"They still haven't succeeded." He grins childishly, but quickly turns serious. "I don't allow them access to the development side, of course." She nods at this. "They are trying like crazy, you understand? It's a matter of pride."

"I understand completely." She continues to nod. She *does* understand. "You must be proud. You *should* be proud."

"It's the work of the whole company. We've got some fantastic new recruits," he says, graciously. "On all teams, actually. Smart as hell."

"I guess the delicate political situation leads to better appreciation of your products." And to more money, she thinks, but does not say. "You no longer lose the best and the brightest to the entertainment industry."

"That's part of it, I imagine."

"And diamonds are made under pressure." She smiles. He nods.

A waiter appears and asks about dessert. Celia indulges herself and Victor encourages it. They both order coffee. The conversation remains light for a while. After dessert, Celia turns serious again.

"So—my second question is about protection of manufacturing and operating details—IP as well as true secrets."

"That is a very sensitive subject." Victor says, neutrally. "I don't know what I'll be able to tell you."

"We've picked Huang Shields as the focus of certain Response Modes simulations," she explains pleasantly, "primarily because your shields are so widely used. I understand sales have more than doubled over the past two years."

Victor furrows his brow. He might be trying to remember when the latest figures went public, although he does not pursue this. "As you mentioned before," he says, his voice crisp, "the country is in a delicate situation. As a consequence, the interest in owning shields has gone up. We are the largest developer and manufacturer of shields, so…"

"The company is thriving."

"Yes. We are doing something important. For everyone."

"I agree. So, my question is this: How much would it affect the negotiations with New Eden and the—shall we say—'perceived readiness for non-negotiated scenarios'—if the relevant people over there had access to technical details of all Huang Shields products—including the newest models?"

"What?" Victor almost jumps out of his chair. "That would never…"

"I didn't mean to imply anything of the sort has happened. It's a question of preparedness, of running through all the possible scenarios, however unlikely. Contingency planning." She adds a delicate smile. "We've been asked to look at this one as it has many complicating human factors."

"Well…" He is still frazzled. She waits for him to recover. "We know they buy or steal any new shields we make," he starts, slowly. "Later on, some of the new designs show up as copies. We see them at the wall." He sighs. "We have to assume they develop zappers and spears in parallel."

"But there is a delay?"

"Yes, of course. A couple of years, I'd say. And they don't always succeed."

"So—with the newest models? What do you expect?"

"I don't think…" His expression hardens. "To be honest, it makes me very uncomfortable to even talk about this. Are you—is anyone—aware of a specific threat?"

"No one has told us anything about a specific threat," she says, neutrally. "We work with hypothetical scenarios." She pauses. "So, *if* they suddenly had all the information, how much would it change?"

Victor scrutinizes Celia—carefully and for a long time. She maintains a normal demeanor throughout. "You're a smart girl," he starts, coolly. "You can work it out for yourself. You don't need me to… and you don't need a bloody 'response model' either!" He stops and draws a few deep breaths. Calm again, he continues. "It would be catastrophic."

"Are you sure? We'd still have the shields. The hypothetical scenario does not involve loss of equipment. Just… information."

"Of course I'm sure!"

Curious glances make their way from the surrounding tables. Victor stares them down. They retreat. The coffee and fancy chocolates arrive and Victor asks for the bill. He is still not looking at Celia.

"Victor," Celia finally says, not exactly pleading, but softly. "I'm sorry if I upset you. I didn't mean to." She stops. After a while, he gives the smallest of shrugs. She continues: "But Huang Shields is so central to everything that's going on now—and to everything that may be going on in the future. We—you—can't ignore any potential weaknesses."

"I know that. It was just the way you…"

"I *am* sorry. I guess I can be a bit insensitive. I offer up my youth and inexperience as excuses." She smiles. He almost does. "Can you forgive me?"

"Nothing to forgive," he says, not sounding very sincere.

"Friends?"

"Friends," he says, turning up one edge of his mouth with a half-sad, half-cynical expression. On the table, she has placed her hand next to his. He pats it lightly and adds "but you are one brazen young lady."

*   *   *

"It's beautiful!" Eiko beams. "Thank you, Celia." Her eyes sparkle.

"You're welcome," Celia says.

"Where did you get it?"

"Oh, I get around." She cocks her head. "Not like you: day in, day out at the lab, doing important work without getting paid for it."

"Contributing—a very small part. And I do get paid."

"Not much, I imagine. Anyway, I had to get it. It just screamed Eiko." Celia fingers the silk scarf that Eiko is holding up. The colors are vibrant and the shapes like giant brush strokes. "So, how do you like this place?" Celia gestures loosely around the room.

"It's amazing—the food, the view, everything... It must be *very* expensive." She opens her eyes wide.

"It's your birthday. Allow me to spoil you. What else is money for?"

"I really do appreciate it." Eiko looks at the scarf again with a happy expression, before hanging it gently over the back of the chair next to her. "Of course, I *did* take you on the trip that lead to your big windfall." She smiles.

"Exactly." Celia smiles, as well.

The waiter interrupts with the next course, which he explains in detail. Their wine glasses are topped up. They toast and concentrate on the food for a while.

"How did you find this place?" Eiko asks at the next intermezzo.

"I didn't tell you, did I? It's the craziest thing. Remember we talked about Leo Huang and his father a couple of months ago?"

"Sure. Victor Huang. His company makes shields. Rafi's parents had quite strong opinions about him."

"Victor took me here."

"Victor..." Eiko looks aghast. "But Celia, he's... old."

"He's an interesting person, actually. Anyway, it was work-related. I was promoting Response Modes."

"Work? Here?" Eiko gestures dismissively. "I thought you were more serious than that."

"It *is* serious—very serious. I promise you. I've learnt a lot of things."

"Like what?"

"Nothing I can share. Not yet." She moves her empty plate to the side and leans forward. "Eiko, promise me one thing."

"What?"

"Promise me that you will always be my friend." Celia speaks softly, her expression serious, a bit sad. "Promise me you will trust I have a good reason for what I do—even if others claim differently—even if…"

"Of course I'll always be your friend!" Eiko looks alarmed. "What are you talking about? Why would anyone-"

"Never mind, sweetie." Celia shakes her head. "I'm just being sentimental. And daft. It's nothing."

"But-"

"But, no but." Celia smiles briefly and slightly forced. "So—our excursion to the new dome—when we were looking for Ben…"

"Yes?"

"That day—you never told me what was wrong."

Eiko shrugs in place of an answer. She looks down, at her plate. Celia studies her friend for a while, but does not push her.

"It's become a huge thing, hasn't it?" Celia says casually, breaking the silence. "The domes. Have you seen the feeds?"

"You can hardly avoid them. It's not from that dome, though."

Celia notices an edge of hurt or disappointment in Eiko's voice, but does not understand what triggered it. She decides not to pursue it. "People talk about those babies like they really know them," she says, instead. "Everyone has an opinion about the names. For once, they are eager to vote."

"I suppose it's not a bad thing." Eiko's voice seems back to normal. "They are little rays of sunshine and hope, aren't they? Pure spring babies. Literally." She smiles. "Well, for a while at least, until…"

"Until what?"

"It's just… what's allowed now and what's not… But let's not talk about that."

"OK."

The waiter returns with the next course, serving as a convenient distraction.

"I was wondering…" Celia says, between bites. "Who are the fathers? For many of the dome babies we only see the mother."

"No one knows," Eiko says, a bit too fast.

"No one knows anything? At the Reserve?" Celia frowns. "But wouldn't it help to know whether they are unedited, like the mothers, or edited, like us. If they are edited, wouldn't that suggest-"

"Why do you ask?" Eiko interrupts her.

Celia waits for a moment before she answers. "I've come across some potentially quite disturbing information." She pauses. "It has been suggested that men from our side may have been... helping themselves. Somehow they have gotten past the wall and..."

"You mean like—raids?" Eiko asks. Celia nods. "And rapes?" Celia nods again. "Where did you hear that? You know you can't trust-"

"It wasn't propaganda," Celia says and thinks for a moment. "But, to be honest, I'm not sure how reliable the source is. That's why I wondered if you had any information. Technically, this scenario can be ruled out if-"

"I'm afraid not," Eiko says.

"You don't know or it can't be ruled out?"

"The latter. Many of the new babies have edited fathers."

"So you guys *do* know who the fathers are?"

"No—just their status. The scientists at the Reserve have been told, confidentially. As you said, this information may contribute to understanding and/or overcoming the block," Eiko says, with forced neutrality.

"So it may have happened," Celia says, slowly. "Raids."

"It's a scary thought if it has."

"Very."

They are silent for a moment.

"They might just be border people, you know?" Eiko says. "I've heard little pockets of unregistered people have been living in those woods. From way back."

"Really?"

"They used to keep very much to themselves." She pauses. "I suppose they were hiding from the authorities." She draws a deep breath. "Now everything has changed."

"I guess it has." Celia nods.

They fall silent again.

Suddenly, Eiko's face lights up. "Those woods—trees—that reminds me: Have you seen Rafi recently?"

"No, actually, I haven't." Celia frowns.

"I finally caught up with him, after a seminar. We had a good chat. It seems that he has given up on trees for his thesis project—they are too slow—and he's working on peas. Peas! Imagine that!" She smiles. "But what he's trying to do actually sounds quite neat." Eiko goes on to tell Celia what she knows about Rafi's project. Celia is half-listening, while trawling her mind for the last time she saw him. It won't come into focus.

At one point during dinner, the head chef comes out from the kitchen to do a round of greetings. He seems to remember Celia, who pretends to be flattered, but actually is quite annoyed. Eiko smiles blandly during their exchange, thinking that the chef is being irritatingly dense and Celia strangely prickly, given that she chose the place. It takes them a while to get back to a normal conversation afterwards.

"I'm taking you home in my pod," Celia says when they finally leave the restaurant. The reception area has a large window with an impressive downtown view. She stops there. "No arguing. It's not safe out there any more, not at night."

"But are you fit?"

"To drive, you mean? I'll set it to auto. It's exasperatingly slow, like an old lady on a Sunday, but it'll get us there."

"OK. I appreciate it, Celia. Thanks for…" She flicks the scarf over her shoulder, where it settles like an exotic bird. "Everything. You are the best."

"At some things, I'm pretty damn good," Celia says with wry smile. She links her arm in Eiko's and they turn toward the window to admire the city lights from a safe distance.

\* \* \*

"Pants on, Rafi. Dinner is served." Celia walks across the room with a large flat box in her hand. Two thin boxes, actually, with "pizza" printed along each side. In her other hand, she has a bottle. Raphael takes one last look at the nighttime skyline and does as he is told. He feels relaxed and content. Tonight's lovemaking was the best they've had in a long time. The first round, in the hallway, was fast and fierce; the second round, on the bed, was slow and gentle. Raphael languidly considers whether he likes it best when she wants his body like a hungry beast, pulling, biting and demanding—or when she slowly whispers an ever-growing list of endearments while caressing those special, secret spots. He leaves it at a draw and moves, fully clothed, to the table. He is ravenous. Celia's place is, in most regards, far preferable to his student accommodation. It has a fabulous view, generous space and it is located in a trendy area. He understands why she chose it. But she never has any food around. Next time he will bring something along, just to keep them going. He smiles at the thought.

"Special occasion?" he asks, indicating the bottle of champagne.

"You're here."

"That doesn't have to be… special occasion. We could…" He looks at her face. She is not helping him. He lets it go.

"I also thought we should celebrate," she continues cheerily.

"Celebrate what?"

"How about… you deciding on your thesis subject. Eiko told me."

"Oh—that—that was ages ago." He shrugs. "And I still have a ways to go."

Celia frowns, unsure how to interpret the dismissive tone. "OK, then, how about a major new contract for Response Modes?"

"Really?"

"Yup. Security-related. Very hush-hush." She gives him a wink and starts pouring. The bubbles rush up. She hands him a glass and they share a brief toast.

Raphael gets busy trying to free the first pieces of pizza. He takes one and starts eating. He starts on a second slice immediately after. That boy is always hungry, Celia thinks and smiles. She focuses on the champagne for a few moments before starting on the pizza. They eat in contented silence.

"I am so tired." Celia yawns and leans back in her chair.

"You work too hard."

"I know. And I have uni in the morning. I probably won't retain a word of it."

"Why do you do it?"

"Uni? Or the company?"

"Well, why both? Why drive yourself crazy with deadlines and no sleep?"

"Why indeed? The money? The thrill? Stubbornness?" She stretches luxuriously. "Some days I'm actually tempted to give it all up. Don't get me wrong. I love the fight. I love the victories. But I get sick of proving them right. I'd love to be the 'fuck-you' rebel instead—be lazy and accomplish absolutely nothing. It would serve them right." She shakes her head. "But I can't do it. I tried." She shrugs. "It lasted less than a day."

"Serve who right?"

"My parents, of course."

"What have they got to do with all this?" He gestures loosely into the apartment. He knows Celia has paid for every square inch of it.

"Exactly. Last time I saw them, we had this huge row. They wanted money. They practically demanded it."

"What money?"

"The money I've earned. Well, some of it. They said that I owe them." Her tone is incredulous, with an undercurrent of bitterness. "They made these huge sacrifices and fantastic decisions for me before I was born: They selected the optimal pre-em cell and ordered the best edits. It's thanks to that, they claim, that I'm now doing so well. Conclusion? I owe them, for having me— their daughter."

"Reverse inheritance. That's kind of funny, actually. It's a little like in the very old days when children were expected to-"

"It's not funny at all! They claim they 'made' me."

"Well, they *are* your parents. So they sort of did, didn't they?" Raphael's tone is gently teasing. He seems to be enjoying himself.

"But to think that they deserve special credit for that—what is *wrong* with them?" She bristles with irritation. "*I* deserve credit for becoming a well-functioning adult after *that* childhood."

His demeanor changes instantly. "They were abusive?" He sounds concerned, even protective.

"No, nothing as thrilling as that. It was a just a mental vacuum. Brains on hold for a constant assault of streaming: stupid flats and worse holos, canned laughter and false suspense. I'm sure that stuff is specifically designed to make unthinking zombies of us all." She makes a face of exaggerated disbelief. "They didn't make any effort."

"Aren't you being a bit unreasonable?"

"Remember when we went to that transfer center a couple of years ago? The New Eden place? That was the first time I saw a paper book. I dropped a really old one and tore a page; I hadn't expected it to be so heavy. I was so embarrassed. You all…"

"That can't be true. In school, we…"

"You and Eiko and Ben… I didn't start out at the same school, remember?"

The hurt in her voice makes no sense to Raphael. He tries for a lighter tone. "How could I possibly forget?" he says. "The day you arrived will be in my memory forever. We were, what, twelve? I took one look at you and I was toast." He moves his chair closer to hers. "Who would have thought?" He starts nibbling on her neck, leaving a greasy, reddish mark.

"Stop that, Rafi." She pulls away from him and wipes her neck with the back of her hand. "I'm trying to say something serious here." She frowns. "My parents don't know me at all. I've always felt that. They never encouraged anything. They don't have a clue what I think or what I want. Still, they act like I belong to them. I'm an investment that has panned out. Time to cash in."

"Slavery was abolished some time ago," he says without humor, possibly a bit miffed, and moves his chair back to where it was. "Anyway, IVF and editing is free, so any claim about huge sacrifices is pure bollocks."

"Screening of four pre-ems followed by four edits—that's free."

"I knew there was a limit, but-"

"They did more than that. They screened more pre-ems for a good starting point and asked for additional edits. It cost them."

"I knew it! You're a hyper!" He immediately looks embarrassed by his outburst. "I mean…"

"Hyper. Hyper-select." She stares at him with distaste. "I see that you've picked up the new vocabulary. Nice." He does not respond directly, or look at her. "It was perfectly legal," she adds.

"But it can go too far, don't you think?"

"It's not like there's an actual recipe for success," she says coolly, "a perfect profile."

"But you *can* stack the cards, can't you?"

"What? You're complaining? The super-privileged kid from the best part of town?" Her eyes narrow with intent. "Your parents are highly educated, financially secure and, to top it off, really nice. They couldn't be better connected. Mother minister of science-"

"Not any more. She-"

"Father professor of physics-" Celia continues.

"Perfectly placed to steer François—"

"Oh shut up about François, will you? I'm talking about you. You're saying the cards weren't stacked in *your* favor?"

"I'm talking about our DNA—the choices our parents made. They did everything for François. Everything! Optimized analytical skills, math skills—full genius profile. So, no, I won't shut up about him."

"You know there's no such thing, right? No genius profile."

"But you can try."

"You can… Never mind. So that's all that matters, is it? DNA?"

"It's a big part of it."

"And your parents did nothing for you?"

"My grandparents put in HQV resistance and had the most obvious disease-associated alleles corrected. A couple more had been worked out since their time—plus cooperative effects. My parents corrected those. But that's all. No optimization, no extras."

"You didn't need it, obviously."

"They did it for François. They didn't for me. That's a fact." He inhales sharply. "They saved second-best for me." He exhales.

"Not that again, Rafi," she says mildly. "Stop feeling sorry for yourself, will you?" She leans forward and moves her hand as if to stroke his cheek. He recoils. "It's just plain silly," she continues. "You're perfect as you are, without any tweaks. You can do anything you set your mind to."

"I could do even more, even better, if—"

"Rafi, stop it! You are smart, healthy, handsome—"

"So you chose me for my looks?" He asks, sarcastically. "Is that it?"

"Sure I did." She moves closer. "Those wonderful brown eyes, those seductive long lashes and, oh, that killer smile."

"Looks don't matter," he says tartly, "for a man".

"You think not? You think people are happy to sit across from Mister Fat and Ugly if they could have you? Look at any movie, any news feed. What do you see?"

"I have no desire to be a movie star. Or a politician."

"But loved? You want to be loved?"

"Everyone does. That's beside the point."

"Really?"

"Anyway, you've got the looks, too. You use it to get-"

"So you admit it. Good looks can be useful."

"For women."

"For women *and* men. You are being willfully stupid, Rafi."

"I'm just living up to expectations," he says, sarcastic again. "The predicted IQ range for my sequence is 120 to 130. With just two edits it could have been 130 to 140. That's a hell of a difference in the real world."

"Not if it's 130."

"What's yours?"

"Now that's a very… *private* question," she says coyly.

He keeps staring at her, angry and insistent.

"Predicted or actual?"

"Both."

"Predicted is 145 to 160."

"What?" He exclaims. He takes a deep breath. "There, you see?" His voice fades.

"Apparently, one pre-em was miles better than the others. They hit jackpot."

"And your actual?"

"You first."

"I didn't take the test. I don't believe in it, actually. It's too formulaic." He shrugs. "So, what's yours?"

"I didn't take the test either."

"I don't believe you."

"Why not?"

He doesn't answer. He stares at the table where the last pieces of pizza lie long forgotten and probably cold.

She pours more champagne. The bubbles are sluggish. "You know the predicted range is only a high likelihood interval, don't you? You could still…"

"Be your equal?" he says, hurt.

"Rafi—come on," she tries. He does not look up. "We are simply different, you and I. Don't you see? We're not interested in the same things. We don't do the same things. Even your brother… Physics, isn't it? He got all those edits you wanted. But he works in another area. What he accomplishes, or not, has no bearing on what you do."

"He's already getting quite a reputation in his field," Raphael says, morosely.

"Arrrgh!" Celia bolts from her chair and spins around to face him. "But is he happy, Rafi? Is François happy?"

"Who knows? You've asked me that before."

"And?"

"It seems you can't edit for happiness." He finally looks up at her.

She sits back down. "Maybe there should be an optimization profile for it: High Probability for Happiness, HPH."

"That should be popular."

"You'd think! Can you imagine prospective parents *not* choosing that profile—and the crap they would face from their kids once they found out?"

He serves up a droll smile. "They'd sue for child abuse—psychological torture."

"I once heard that Down's syndrome kids—you know, one of those mental retardation syndromes we eliminated ages ago—that they were really happy kids." She immediately regrets the comment.

"Fuck that," he says. "Slightly stupid, but blissfully happy. Would you seriously want that?"

"You mean, like, have a lobotomy? Of course not."

"There, you see?" He perks up. "Happiness is just an overrated fudge factor—vague and mushy. Anyone who says they can measure it or predict it is full of shit." She shrugs. "The truth is, we all want the same attributes," he continues with more edge. "Eons ago it was strength. Now it's mental superiority—and, I suppose, pleasing looks. Those of you who had it all handed to you in your DNA like to pretend the tweaks are not so important. But that's easy to say when you've got them." He looks triumphant.

"Rafi—this is hopeless." She looks sad. "You seem stuck on proving everyone right. Having seen your sequence, you resign yourself to self-indulgent imperfection. We're *all* imperfect. The human race is a pitiful bunch. Maybe it's a good thing that we can't reproduce any more."

"Don't say that! Not even as a joke."

"Oh, I don't mean it. Or maybe I do." She throws up her hands. "Whatever. My opinion is irrelevant, anyway. You're the one who could do something about it. But you won't even try."

"What? Me? They've already got the best brains working on it. What could someone like me add?"

"Someone not optimally edited, you mean? You're so predictable. Always the same excuse."

"And, as it happens, you don't solve scientific problems by simply deciding you want to solve them. Scientific discovery doesn't work like that."

"How *does* it work?"

"You need to put in the time and the effort, of course, but you also need luck and… inspiration and…—things you can't force."

"So people don't inspire you? Plants do?"

"It's not just that. In our field there's room to breathe. If you have an idea, you can follow it through."

"And if you worked on the fertility block, you couldn't?"

"It's super competitive. Good ideas would be appropriated in a heartbeat."

"And that's why you're not working on it? The competition?"

"No. It's because I like what I do. And, as I said, all the brightest minds are already working on it."

"Like your precious brother, François."

"No," he says, sounding exasperated, "he's a physicist. He might be good for a bomb or two, but not for this."

"I know that," Celia grumbles, "I was just…" She sighs audibly. "Eiko tries, at least. She's got a lot more to be upset about than either of us, but she's doing what she can."

"What is she doing, exactly?"

Celia looks at him with surprise. "I thought you'd know that. I took her out for a fancy birthday dinner last week. That's when she told me about your thesis project. I assumed the two of you also talked about her work."

"We didn't get to that, actually."

"She told me a bit about it, but I'm afraid I didn't really get it. You'll have to ask her."

"She's only a PhD student," he says, somewhat sourly. "They wouldn't have her doing anything critical for-"

"Don't!" Celia snaps. "Don't you dare!" She stares at him, hard. "Wallow in self-pity if you have to. Leave the important stuff to everyone else. But don't… I'm so tired of you always…"

The words hang in the air for a while, unfinished but also unchallenged.

Celia empties her glass, not noticing what she is drinking.

"Why are you so angry?" he asks.

"Me? This is not about me."

"No?"

"No. We were talking about you. We were talking about Eiko—and about whatever is happening out there." She gestures loosely toward the windows. She sighs. "But I guess we've said all there is to say." She stands up. "I'd like you to leave now. I need sleep. I have a lot of work to do tomorrow."

Raphael looks like he has been slapped.

"Leave?"

"Yes."

He grabs the bottle, pours and drinks. He pours again. He remains seated. She remains standing.

"So," he says, "are you going to see your sugar-daddy tomorrow?"

"My *what*?"

"Don't bother trying to deny it, Cee. I've seen you with him—Huang senior. Distinguished-looking, I suppose, if you are into that kind of thing. He's certainly rich as hell. Powerful too."

"What do you mean you've seen me? You've been following me?"

"Of course not! I was just walking past the… So, you *are* seeing him."

"He's a business contact!" She throws up her hands. "Powerful? Yes. Useful? Absolutely. Stop this nonsense, Rafi."

"Are you two…?"

"No! He's far too old." She tries a smile. "Plus, I have you for that."

"*Have* me?"

"Rafi, please, that was a joke." She touches his arm lightly. "But I really *do* have a long day tomorrow. Finish that-" she indicates his glass "and then you have to go."

A few moments later, Raphael picks up his jacket and they walk to the door. He stops up a couple of steps short and says, pensively. "I guess you'll never understand, will you? You don't know what it's like to have your future snatched away."

"Snatched away? What are you talking about?"

"The year after you joined our class, we got split up for some subjects— maths, science and computing."

"I remember. I never liked the way they did that. All those tests—and for what? Luckily, Eiko and I were able to-"

"You took my spot."

"I did what?"

"In the A-prime stream. You jumped the queue."

"There was no queue, just a bunch of tests."

"You know what I mean."

"There were ten of us."

"The year before…"

"I get it." She sighs.

"If you hadn't had all that extra… it would have been fairer."

"Nothing is fair, Rafi," she says, sounding genuinely tired. "Don't be such a baby, OK? Now go." She kisses him quickly on the cheek and closes the door.

The autumnal air is crisp and pleasant, but Raphael does not notice this. While he walks, and while he waits for the night shuttle to come around, he keeps reliving that terrible day, more than nine years ago, when he wanted to end it all. Again and again, he sits alone in a room, reading that devastating notice "Welcome to the A-stream". He knew what it meant: the almost truly gifted.

He gets on board the driverless shuttle, flicks his ID past the reader and lets himself fall sluggishly into a blue plastic seat. He turns his head and looks at the wallscreen for distraction. The sound is off, but it is obviously a news-feed. With the backdrop of some government buildings, he sees the now familiar sight: a mob of ordinary people shouting, shoving and generally being quite uncivil. Angry men and women with linked arms form the outer rim of the crowd. Their coordinated flash-shirts spell out slogans. The gradual buildup of letters keeps you reading and guessing. Cheap trick, he thinks to himself. CDL, no doubt. Stupidity whipping up yet more outrage and fear. How do they manage to get it so wrong? New Eden did not cause this mess. Sure, they are milking it as much as they can, but... He reads the slogan. This is not CDL. This is new.

*They take your job*
*They jump the queue*
*Oust the hypers now*

He looks around quickly. There is no one nearby to have read his mind. He breathes in and out, slowly. He feels just a little bit sick to the stomach.

# 9

## Father Marius

"Father says it's time to go," Anne mumbles, not quite looking at Ben. She is standing in the doorway to the courtyard. A gust of wind blows fallen leaves into the kitchen. The outside air is cool and damp.

"Tell him I'm not coming," Ben says.

"But, he says…" She gives up quickly and disappears from sight. The door remains open, latched thus from early morning. Ben cannot bear it closed.

"I sure don't need any of that crap," he says to the pot in his left hand. A crust of age-old burnt residues lines the side of it. He starts scrubbing as hard as he can while cursing under his breath and keeps going until he notices Milly in the corridor behind the kitchen. She has her coat on but not buttoned. "Do *you* want to go to church?" he asks her, almost aggressively. She shakes her head and stays put, her eyes on his hands.

A few moments later, the light in the kitchen dips. This time, it is Harold in the doorway, filling it with his bulk. "Benjamin," he says, "leave that for tomorrow." He nods at the pots and pans piled up by the sink. "You should come with us. Now, please."

"No!" Ben stops scrubbing and looks at Harold, defiantly.

"I know things are difficult right now." Harold's voice is strained. "But all the more reason for the family to… pull together. You should give the new priest a chance. He has asked-"

"He keeps track of who does and doesn't show. I know." Ben starts scrubbing again. "I don't care. Let him damn me to hell."

"I will *not* have that kind of talk in my house." Harold says, his face coloring. "Mathilda was my sister. Her passing has affected all of us." He takes a deep breath. "God has his reasons. We cannot presume to know-"

© Springer Nature Switzerland AG 2020

P. Rørth, *The Unedited*, Science and Fiction, https://doi.org/10.1007/978-3-030-34624-9_9

"Only a sadistic, cruel God would make Tilly suffer like she did. How can you-"

"That's enough!" Harold interrupts him, his voice finally unconstrained. "Go to the shed. Now!"

"No. I am not a child."

"You are living under my roof." The veins on Harold's neck are throbbing above the stiff white collar. "As long as you are, you will obey my rules." He steps aside and points to the narrow door across the courtyard. "Go. And stay there until I return." Ben puts the pot down noisily and throws the scrubbing brush into the sink. He will not give Harold the satisfaction of seeing him hesitate. Peter is afraid of the shed, the household's designated space for imposed contemplation. He will never admit it, but he is. Ben is not afraid. He walks straight across the courtyard, goes inside and closes the door behind him. Harold follows with long, swift steps, and noisily slides the two deadbolts into place. He turns, re-crosses the courtyard and joins his family, waiting stiffly in the driveway. They have watched, but not heard, the exchange. No one asks for details. They move off, silently.

A few minutes later, Milly emerges from the kitchen. She looks around quickly, then runs across courtyard and knocks three times on the bolted door. The knocks are mirrored by Ben. She slides the lower deadbolt back easily, but has to fetch her stool to reach the upper one. Light floods the dark and damp shed. Ben has to close his eyes. He extends his arm in the direction of the door. Milly takes his hand and leads him outside. Once Ben's eyes have adjusted, he notices the many scratches and dents on the inner surface of the door. He squeezes Milly's hand and they walk back to the kitchen together. Wordlessly, Ben continues scrubbing and cleaning. Milly helps him as best she can. Ben is well aware that this is not suitable work for a Sunday. Tilly would not approve. He feels simultaneously proud and ashamed of his compulsive protest.

Daisy, Peter and Anne return first. Daisy goes to the kitchen. If she notices how clean everything is, she does not say. She simply tells Milly to get the coffee and cake ready.

Peter kicks the door to the shed as he passes it: once, twice and a third time, much harder.

"Stop that," Ben growls from within.

"Go back to where you came from, you witch-loving bastard." Peter says and secures the upper deadbolt.

Harold arrives half an hour later. He goes directly to the shed and unbolts the door. With a calm and steady voice, he asks Ben to please join him for a conversation. He turns and walks toward the front door. Halfway there, he changes course and heads for the courtyard bench instead. Despite the damp, he sits. He gestures for Ben to do the same. Ben remains standing, right across from him.

"I am sorry I lost my temper earlier," Harold says. "I know you had gotten very attached to Mathilda." He looks at Ben's face. Ben looks at the ground. They remain like this for a short while, in silence. "Father Paul told me he noticed your distress at the funeral," Harold continues. "He thought… He understands and respects your sorrow. He asked me to do the same."

"Father Paul did not know Tilly."

"Not for long. That's true. But he is trying to help."

"He just wants me to show up for church."

"We all do. The congregation is there for you."

"Bullshit. They want me gone. You do too, now that Tilly…" He stops talking.

"I know you grew up without God's word, so it is hard for you to accept-"

"I'm an adult," Ben says. "I can think for myself. I simply don't believe that…"

Harold takes a deep breath. "You do not believe. That saddens me, of course. But you could at least-"

"You don't want me to make you look bad. I understand completely." Ben's tone remains harsh and uncompromising.

"You were the one who decided to come here," Harold says, curtly. "I agreed to take you in and to give you a home. All I ask in return is that you show some respect."

"Tilly took me in. She cared. You don't."

"Mathilda was a very special person. But I still have a household to run and a family to worry about."

"You want me to leave."

"God tells us we must have patience." Harold sounds like he is losing his. He rises from the bench. "God loves Mathilda just like-"

"Tilly is dead!" Ben shouts as he takes a step toward Harold. "*Your* God did that! *Your* God made her suffer-"

Harold's hand flies out and a slap lands on Ben's right cheek. The smack is unexpectedly loud. Ben's head turns with the impact, but he does not utter a sound. Two seconds later, he readjusts his head with deliberation.

"I can't…" Harold breathes deeply before he continues. "You sound just like her, do you know that? The pride, the arrogance, always so sure you are right and everyone else is just plain stupid." His voice is low and tense. "If only you were a little bit more like Mathilda and a little bit less like your pig-headed mother." Ben looks at the ground and does not respond. Harold steps even closer to Ben and continues talking with the same quiet intensity. "Do you have *any* idea what it did to this family that she just… ran off? Sneaking away in middle of the night… She never thought of anyone but herself. She disrespected her father, her family, the church—everything. Mathilda was heartbroken when she left. Did you know that?" He stares at Ben, long and hard. "I was here and I know. I saw what happened. Mathilda may have forgiven your mother. She may have felt that you were somehow…" He throws up his hands. "I don't know… But we are not all saints." He no longer sounds angry. He sounds deflated, even sad. "I want you to…" He does not continue. Instead, he shakes his head and walks slowly toward the front door. The lights have been turned on inside. Anne and Peter are standing by one of the living room windows, but are no longer looking at Ben or at their father. Ben turns and walks toward the kitchen door.

\*   \*   \*

"Hello, Ben." A deep and mellow voice breaks the evening quiet. Ben looks up from his book, startled. "Harold said I might find you here." The man is standing in the doorway between the kitchen and the main part of the house. He is wearing a long, dark cloak, but his head is uncovered. He steps forward and closes the door behind him. Except for Ben's reading light and diffuse light coming from the rear corridor, the kitchen is dark. "My name is Father Marius," the man continues as he moves closer. When he reaches the end of the long bench, he stops. He sits down and some of Ben's light falls on his head. His face is roundish and slightly flushed—soft, somehow, with pale eyes and eyebrows. The light catches traces of red in his hair and beard, along with some gray. Ben thinks he may have seen him before, but he is not sure. "What are you reading?" Father Marius continues, lightly.

"Nothing," Ben says, closing the book and placing it on a nearby stool. "Just a story." He folds his hands and waits. Father Marius smiles sadly. Ben senses that his interest is genuine, not overbearing. At this point, he also recognizes him, both the name and the face, and therefore relaxes somewhat. "I saw you at Tilly's funeral," he says. The visitor's head had been covered much of the time, but Ben had caught a glimpse of his face during the service. He had been visibly distraught.

"Tilly was a close friend of mine, way back." Father Marius sighs. "She was a wonderful person, generous and caring… But you know that, I'm sure." He pauses. Ben nods. "She will be missed," Father Marius adds and then looks away. After a while, he looks at Ben again. "I grew up not far from here. That's how I got to know Tilly—school and so on." He tries to smile. "Did she ever mention me?"

"She did…" Ben says. "Father Len mentioned you as well."

"Of course, Father Len. We met at the academy and became friends. I was happy he got posted here."

"I liked Father Len very much. He listened and he…" Ben's face contorts slightly. "How is he? Is he OK?"

"He is…" Father Marius hesitates. "He's settling into his new posting, I think." He looks at Ben again. "We don't get to choose, you know. He would have stayed here if he could."

Ben nods. "Tilly and Father Len said you helped look for my parents."

"I did what I could. I'm sorry it didn't-"

"It's not your fault," Ben says, his voice hardening somewhat. "I know they're gone. There was an accident and they died. That must have been the truth all along."

They sit without speaking for a while.

"You have been through a lot, Ben," Father Marius says softly. "First losing your parents, then leaving everything you have ever known and coming here, to such a different life. And now… Tilly." He pauses. "It must seem so unfair. I understand why you are angry."

"I'm not angry," Ben says, too emphatically. "I'm just…" He blinks rapidly. "I don't know," he finally says. "And I don't know what to do."

"I really *do* understand." Father Marius moves his hand forward, as if to place it on Ben's arm, but pulls it back again. "I have a suggestion," he continues. "Why don't you come with me? At least for a while?"

"With you?" Ben frowns. "Why? Where to?"

"I'll be honest with you. I've already talked with Harold and he thinks it's a good idea. We both think you might need a different… context. Away from…" he gestures around the kitchen "all the sadness. You deserve a chance to build yourself a life, Ben." He searches Ben face. "Tilly would agree."

"I…" Ben glances toward the rear corridor. "I'm fine," he says, firmly. "I just miss her, that's all." He shakes his head. "Anyway, I can't go with you. I don't believe, you see." He looks at Father Marius, but avoids his direct gaze. "It was fine with Father Len. He talked about faith, sometimes, but he accepted that I was—that I did not believe—do not believe—in God. So I can't go to the Church. I just can't."

"The Church?" Father Marius looks surprised. "No, no… I was going to take you to my mother's place."

"Your mother? But why?"

"Obviously, she's no longer young. But she's quite feisty—and fond of strays." He smiles. "She has a small farm. It used to be her family's farm, you see. When my…" He stops and shakes his head. "She's a bit like Tilly, actually. You'd like her."

"But why?"

"She could use the help. And the farm is…. Well, it's in a beautiful area. I come around whenever I can, but I have my duties to attend to, so it…" He hurries ahead. "Don't worry, it's not too isolated and the farm work is not too arduous." He smiles again. "You'd have plenty of time for reading." He nods at the book on the stool. "We have quite a lot of books there. Tilly would approve."

"But what about here? And the school? I…"

"You can't go back to the school, Ben." He sighs. "The parents know about your views. Harold felt he had to…"

"Harold felt he had to, did he?" Ben regrets the words and the tone immediately. He sighs and lets his shoulders sink.

"The parents, they… they worry. They knew Tilly, of course, and they loved her—everyone did. But you are…"

"Different."

"Yes." The answer sits solidly between them. "And the family, Harold assures me, they can manage here. His wife and daughter, plus the girl-"

"You mean Milly?"

"I don't know her name. The little mute girl who Tilly was taking care of."

"Milly. I can't go anywhere without Milly," Ben says, firmly.

"But…" Father Marius looks baffled. "The girl has to stay with Harold and Daisy."

"Why?"

"It's complicated."

"Explain."

"She's… I don't know all the details. Tilly told me once, but… I think that was the arrangement she made with the girl's mother."

"Her mother? I thought Milly was an orphan."

"Her mother gave her up. It's what normally happens in these situations."

"What situations?"

"Don't be dense," Father Marius says, suddenly irritable. "I'm sure you understand. A girl gets pregnant. She is not married. Both their lives are ruined, if they don't… make arrangements. It works out, generally."

"The children are sold off as servants?"

"Gracious, no." Father Marius sends Ben another baffled look. "Is that what you think…? No. The children are adopted. Tilly petitioned to adopt, I think. Apparently, she knew the mother. But it was considered better that Harold and Daisy adopt her, since they are a married couple."

"But they don't care one bit about Milly. To them, she's just a servant."

"We can't control that. The family decides how best to care for the adoptee. That's just how it works."

"Then I'll stay and take care of her. Without Tilly… Milly needs me. Tell Harold I won't be causing any more trouble. I'll…" Ben looks unsure.

Father Marius seems to notice the hesitation. He waits. "Ben?" he finally asks. "Are you sure this is what Tilly would want? You meant a lot to her. I could tell that from-"

"If you really knew Tilly," Ben says with bitterness, "you would know how important Milly was to her. She wouldn't want me to leave her here—with them." He gestures at the closed door with an expression of distaste. "You keep saying that you were close to Tilly. Why should I believe you? You just show up at the funeral and-"

"I was going to marry her," Father Marius says, with a sigh. "She was the love of my life. Did she tell you that?"

Ben shakes his head. "But you didn't," he adds, after a bit.

"Do you want to know what happened?"

Ben holds off for a while, still upset. But he is also curious. "Tell me."

"I was born in Paxton, a few miles west of here. The school in Hatton was convenient, and that's where I met Tilly. I knew your mother as well, of course—the pretty and popular sister. But I loved Tilly. I have always loved her. She was the sweetest, sunniest girl. She made the whole world a wonderful place." He smiles at the memory, briefly. "I asked her to marry me the day she turned eighteen. I had planned it for months. We were at our favorite picnic spot, just the two of us. I had put flowers everywhere." Another brief smile. "And she said yes. She laughed, happily, and said yes." He stops. "She said yes, but then…" He sighs. "A couple of months later, she told me she couldn't marry me after all. She cried and cried and almost couldn't speak, but when she did, that was what she said. I tried to give her space. I offered her more time. She didn't want it. She told me she didn't love me. She had fallen in love with someone else."

"My father," Ben says, quietly. "Jacob. She told me about it, before she…"

"Right. Jacob. The charming mystery man swoops in and…" Father Marius says bitterly. "He was a real-"

"He was a *great* father!" Ben exclaims. He looks surprised at himself.

Father Marius shifts backwards. "I'm sorry, Ben," he says, looking chastised. "The truth is, I never knew him all that well." He is quiet for a while. "I'll tell you the rest of the story, even though it is quite unflattering. I suppose I… Well, I was young and angry and stupid. I saw what happened when Jacob started coming around. He was fun, he laughed, he made up games—he danced. They both fell for it, Eleanor *and* Tilly. I was so jealous. It was eating me up. I admit it. I was stupid."

"You said that already," Ben grumbles.

"So I started to flirt with Eleanor, to make Tilly jealous."

"But she…"

"We were all young. She was game for a bit of mischief—and probably had her own agenda." He sighs. "It was a mistake. I realized that, pretty quickly. She must have, as well. They disappeared soon after, Eleanor and Jacob. Gone forever." He takes a deep breath. "Tilly was devastated. She turned away from me. I never knew how much she had found out. So I left."

"You became a priest."

"It seemed the best thing to do. It meant leaving the area, studying hard, focusing on other people and their concerns, being useful. Then my father died and my mother moved back to where she grew up. There wasn't much left of the old farm, but… I go see her as much as I can."

"And Tilly?"

"She got in touch sometime after Father Len was posted here. She must have found out that he and I had become friends. It was just a few letters, greetings passed on, everything through Father Len. She insisted that I should not come back. It would be too difficult, she said. I respected that." He looks Ben straight in the face. "If I had known how sick she was, I'd have come back no matter what she said. I had no idea."

"Father Len didn't tell you? He knew."

"Apparently Tilly told him not to. He respected her wishes." He sighs. "Too much respect and not enough courage, I'm afraid."

"The first cancer was seven years ago, she told me."

"I didn't know."

"They cut it out. But it came back."

"Yes…"

Ben looks toward the dimly lit corridor for a moment. "She wasn't afraid of dying, I don't think. At least it didn't seem that way. Some days, she said she knew there was something benevolent out there, behind it all. God, I suppose. Other days, she was less sure."

"I wish I could have been there."

"She and Father Len talked quite a lot."

"That's good."

"But the pain… In the beginning, she tried to pretend it wasn't serious—just a cramp or something. But after a while…" Ben breathes deeply. "When it got really bad, her face would go all stiff, her eyes darting around, desperate, her grip so hard… finally she would whimper or cry out. She tried not to, but once she started…" Father Marius nods, but does not interrupt. "I had to take Milly for walks, sometimes. Tilly insisted. When I came back, she'd be lying there all spent and weak—and full of apologies." He blinks, shakes his head. "At night, I'd watch over her while I got the next batch of Margot's extract ready."

"You got herbs from Margot?"

"Yes, for the pain. It was the only thing that helped." Ben looks carefully at Father Marius. "I really liked Margot," he continues. "I only met her that once, when I went to get the herbs, but we talked for several hours. She asked lots of questions, and she listened. She didn't seem to be afraid of anything, even though… She was very old, I think."

"She *is* very old," Father Marius says, nodding slowly. "Amazing, really, when you think about it."

"You know Margot?" Ben asks, incredulous. Then he remembers that Father Len did, as well. He grabs on to the next thought. "You said she *is* very old."

"Is. She's been living with my mother for the past couple of months. An old lady taking care of an ancient one." He smiles with warmth and mild wonder. "They are quite a pair."

"But… Her cottage was broken into—ransacked—the kitchen was a complete mess. I thought…"

"Yes, I heard. It happens whenever the fanatics get too fired up and need to take it out on someone. But she heard them coming and slipped away. The thugs were probably relieved she wasn't there. No one really wants to hurt an old lady."

"Even if she is… a witch?"

"A witch?" Father Marius laughs indulgently. "You really *do* think we live in the dark ages, don't you?"

"No, I just… that's what people in the village say." Ben seems put out.

"It's alright, Ben." Father Marius smiles. "People say all kinds of stupid things." He leans forward and catches Ben's eyes. "Are you sure I can't convince you to come with me? Now that you know Margot is at Birch farm as well? Margot and my mother would both be thrilled to have you, I'm sure."

"Birch?"

"My mother's maiden name. I have to say that-"

"So you know Vera Birch? Who used to live with Margot?"

"I knew Vera many years ago." He pauses. "She's my mother's… adopted sister, I guess it is. And you're right. She *did* live with Margot for a few years. She was a very unusual person. Apparently she-"

"Crossed over to my world. Became Vera Weiss? The famous Dr. Vera Weiss?"

"You knew my aunt Vera?"

"I used to call her aunt Vera."

"*Your* aunt Vera?"

"She wasn't really my aunt. We just called her that. She came to the house a lot. I liked her." Ben smiles. "She was also our doctor."

"That's nice."

"She died some years ago," Ben says, quietly.

"I know. I haven't told my mother. I should have, but…"

"I told Margot."

"Oh…"

They do not speak for a short while.

"She was kind of famous for what she did," Ben says, "and very critical of New Eden. Wasn't that a problem for you? For your position?"

"They never seemed to make the Weiss-Birch connection. And my surname is Haug, anyway. I had a moment of soul-searching when I filled out the background questionnaire, but I figured…" Father Marius shrugs.

"Crazy," Ben says and grins. "Aunt Vera helped my parents with all kinds of things, from when they first came over." His face darkens. "I think."

"I believe she helped a lot of people." Father Marius hesitates. "Whatever their circumstances."

"There were other illegals?"

"I don't know the details, to be honest. They didn't want me to know. To protect me, I suppose. I believe the authorities over there got suspicious at one point. They tried to…" He doesn't finish the thought. "Of course, everything has changed now. They must wish Aunt Vera's underground railway was still flourishing."

"Everything has changed? What do you mean?"

"You don't know?"

"Know what?"

Father Marius straightens up. He looks at Ben for a while. "Ben," he says with some intensity, "when did you come to New Eden?"

"Two and a half years ago."

Father Marius gets up from the bench and starts pacing the room. He stops once by the door he came in through, checks that it is properly closed and listens for a while. He moves on. Ben looks at his circling form, more and more curious.

"So, my aunt Vera," Father Marius starts, "you said she was famous. Was that for crossing over? For criticizing New Eden?"

"Well, that too. But also because she… She was a doctor and she had cancer."

"Yes, I know. Despite all the years of fixing—somatic editing I think you call it—she still got it."

"She also did cancer research, and, according to a friend of mine, when she got sick, she started donating cells to research. Do you know about cell lines and that kind of stuff?"

"My family is full of doctors, of one sort or another. I can't completely avoid it."

"My friend even had Aunt Vera's genome sequence on file, from her work with the cell lines. She was thrilled that I knew the famous Dr. Weiss in real life." He smiles, remembering the conversation.

"Your friend, who is she? What does she do?"

"Her name is Eiko Carr. She's a PhD student at the Reserve. At least I think she is. When I left she was just about to…" Ben drifts off. Father Marius stops his pacing and pulls up a chair, across from Ben's.

"A PhD student at?"

"The Reserve. The reproductive services. Her parents work there as well: Paul Carr and Yuriko… Ito, I think. He's a research scientist, she's a clinician."

"So you were studying with Eiko? Were you also at the Reserve?"

"No, we went to school together. I hadn't quite figured out what I wanted to do. Everyone else…" He frowns and restarts. "Another old friend of mine is studying with Eiko, biology and biological engineering. His name is Raphael Delacroix Winter." Father Marius is looking at Ben very intently, focusing on every word. Ben senses that what he is saying is important, but does not quite understand why. He is happy to continue. "Rafi is the son of the minister of science, Marie Delacroix. But he's quite cool about it, actually. He-"

"Former minister of science," Father Marius interjects. "The government has changed."

"OK, but…" Ben says, puzzled. "How do you know this?"

"It's no secret. And we have to stay informed."

"You work for the archbishop in Petersburg. Tilly told me."

"I do." Father Marius does not explain further. He rubs his beard. "So," he continues, "Rafi and Eiko, are they… good friends of yours? Close?"

"They are. Well, they were. We've known each other since forever."

"And you know their parents?"

"Of course. I spent a lot of time at Eiko's house when I was a kid. In many ways she was—is—like a sister to me. We were the only ones without…" He drifts off again. Father Marius appears to be thinking. "They might be mad at me now, I suppose," Ben says, a bit saddened. "Eiko, in particular."

"Why?" Father Marius is paying attention again.

"Because I ran off like that. Without explaining or…. It wasn't planned. It just became possible. And I had to find out about my parents."

"Back up a bit, Ben. You ran off, you said. From your friends?"

"We were all going to visit New Eden, together. Everyone was interviewed and so on, but then it turned out that Eiko…" He stops, looks at Father Marius, and restarts. "They weren't allowed to go. I wasn't either, but… Anyway, it was easy, actually. At the transit center there was this cook, Hetty, she-"

"You came in via a transit center? How peculiar. I thought they were only for show."

"It *was* pretty empty, that's true. We wondered about that—why no one else was there. I had forgotten that part."

"So your friends' high-placed parents got you access?"

"No, nothing like that." Ben laughs. "They didn't even know. I bet they were furious afterwards." His smile goes away. "No, that was another friend of mine, Leo. He's a totally amazing hacker. He bumped us to the front of the queue, made sure we couldn't be tracked when we went there—all kinds of stuff."

"This Leo… who is he?"

"He's… We in met in student housing." Ben thinks for a moment. "When we were seventeen or eighteen. Leo Huang."

"Huang?" Father Marius gives a jolt. "Any relation to Victor Huang, of Huang Shields?"

"Leo is his son and…" Ben frowns. "But they never spoke. Leo hates his father. That's why he was in student housing."

"Hmm."

Ben gets up and checks the fire in the stove. It is almost out. He adds another log. The nights are getting cold. He steps into the back corridor and listens. He hears the faint sound of sleeping breath. He leaves the door open and moves slowly back toward Father Marius, who still looks to be deep in thought. Leo, Rafi, Eiko and Celia are all there, inside Ben's head. They chat, argue, joke and laugh. He postpones letting them go for as long as he can.

Finally, Father Marius looks up. "So, you don't know anything about what has happened in the world the past couple of years? Here—or over there? Is that correct?"

"Pretty much. Some time ago I noticed that Peter and his friends were getting all excited—something about the need to defend New Eden. I asked

Father Len about it. All he knew was that there was some tension at the border." Ben grabs the back of his chair and sighs. "Since then I've been preoccupied with everything here…" He gestures into the room behind him.

"Of course. So you haven't heard about the fertility crisis affecting chartered countries? Or about the bilateral negotiations?"

"No! Tell me." Ben sits down quickly.

Father Raphael tells Ben what he knows about the fertility block, which is not everything there is to know, but enough for Ben to go "Holy cow!" followed by "Sorry", followed by several other exclamations.

"So, you are saying that the edited are all sterile?" Ben summarizes.

"It seems so."

"And they haven't figured out why?"

"As far as I know."

"Well…" Ben shakes his head. "That's pretty wild."

"It is."

"And it certainly… changes things."

"It does. The power balance has shifted, that's for sure."

"Bizarre. Who would have thought?" Ben laughs, suddenly, unexpectedly.

"What?"

"I'm unedited, as well."

"I didn't know that. I thought at least—like for Aunt Vera."

"Nope! I am completely—I won't say pure—but yes, unedited."

"How is that possible?"

"It's a long story." Ben shrugs. "I just never imagined it would be an advantage." He gets up from his chair, abruptly. Now he is the one pacing the room while Father Marius is watching.

"Listen, Ben. I haven't told you the most important part."

"No?" Ben sits back down.

"Well, the part I need your help with."

Ben looks surprised. "OK…" he says and leans forward.

"There are various ideas about how to deal with the fertility problem. Not surprisingly, some people are trying to take advantage of the situation. Politically… well, I won't go into that. But an important factor is whether the HQV virus is still around."

"At the transit center, they told me the whole HQV thing had been exaggerated and that it…" Ben thinks for a moment "that it was a thing of the past. I thought he meant it had been eradicated."

"He? Who told you this?"

"I think he was called Father Elias."

"Father Elias… Of course…" Father Marius lifts his hand in a gesture Ben does not understand. "I wish I could tell you I knew that to be true. The thing is… I'm not sure."

"You're not sure whether it has been eradicated or not?"

"Exactly. There are many things I am not privy to, but I do observe—*and* I have both doctors and skeptics in my family." He pauses. "I've never talked to anyone else about this, not directly," he continues in an even softer voice, leaning in close. "There's something here that's… They call it possession. We are told it is God's punishment. Maybe it is. Apparently, the headaches are ferocious. And I think it is infectious. What happens is this: at the first sign of 'possession', the victim is sent to the sanctuary and the rest of the household is quarantined."

"That sounds like…"

"Yes."

They both sit back, each in their own thoughts. Father Marius leans forward again and Ben does the same.

"Do you understand how important this is?"

"I do."

"Will you help me, Ben?"

"I…" Ben closes his eyes, lets his shoulders drop. "Yes. If I can—and if…"

"I understand if you don't feel you can leave with me right now," Father Marius says. "I'll look into arrangements for the girl Milly."

"Thank you."

"I will visit you again soon, hopefully with good news about the girl—and a plan. I will also bring you some samples. I want you to give these samples to your friends back home, or to their parents or colleagues. Give them to people you trust and who know how to find out the truth. They will do the right thing."

"I thought you said the borders were sealed because of all this."

"There is still a crack."

"Like in Aunt Vera's time?"

"It's an even smaller crack now, but we have to try. You can do this, Ben."

"I can?"

"You can. It may be dangerous, though."

"I guessed that." Ben draws a deep breath. "I'm not very brave, you know."

"I think you are," Father Marius says and puts a hand on Ben's shoulder.

"No," Ben says, "but they are my friends."

\* \* \*

Ben stays true to his promise. There are no more outbursts. He does the household chores, his and Tilly's. It keeps him occupied. He even goes to church, eyes downcast. He stays away from the school. Harold praises Father Marius and his good influence. Ben does not disagree.

A few weeks later, Ben is making bread when he hears a scream. The scream is human, high-pitched and desperate. He does not recognize it. He leaves the dough and rushes outside, into the semidarkness of early dusk. He stops up to listen and sees his own breath, expanding in the stillness. Harold comes out through the front door and joins Ben in the courtyard.

"What was that?" Harold asks. "Where did it come from?"

Ben sees movement next to a group of naked trees by the fence. "Over there, I think," he says and raises a white-dusted hand to indicate the trees. They both run over, Ben in front, Harold behind, panting loudly. Ben sees a woman with long, wild hair, trying to lift something off the ground. The something is covered in dirt and is kicking and screaming.

"Milly," Ben shouts, when he gets close. "Milly, it's Ben. I'm over here." She tears herself loose and runs toward him. Ben crouches down and wraps his arms around her small body, hugging her tightly. Harold comes up behind them and runs toward the woman. He grabs her hair. Now she is the one screaming. The scream is not just pain, but also fear, anguish—or madness. Ben cannot tell which.

"You crazy whore!" Harold yells at her. "Go back home." He gives her a forceful shove and she falls onto the hard earth, her coat flying open to reveal a flimsy cotton shift. She picks herself up, grabs hold of the wooden fence and shuffles backwards along it. "Don't ever come near us again." Harold continues.

"Thank you," Ben says to Harold, not really knowing why.

"This little bastard is nothing but trouble," Harold says, indicating Milly with a rough movement. "Born in filth and covered in filth. Bloody Mathilda and her charity projects." He starts back toward the house. Halfway there, he turns to Ben and Milly again and says, in disgust. "I should send her back to where she came from. To that crazy whore…"

Shocked, Ben does not respond. He simply hugs Milly tighter. She is still so slight, so fragile. He picks her up, adding ghostly white marks onto her coating of mud, and walks slowly back to the kitchen.

"*My* Ben," Milly whispers in his ear. Her voice, those two first words—it is so unexpected that he almost drops her. He is not quite sure what just happened. But, for the first time in a very long while, he feels grateful for something.

# 10

## Peas and Politics

Why the hell am I here? Raphael wonders as he tries to make his way through the masses of people. Civic Square is dense to overflowing and the mood is not cheerful. More to the point, why the hell is *she* here? He sees a crumpled cardboard box on the sidewalk. He kicks it. And since when has littering become acceptable? The cardboard hits someone's leg. Raphael looks up and apologizes quickly. Surprisingly, the nearby faces show less anger than he feels. He tells himself to calm down. It works, somewhat. He starts looking around more carefully. The flash-texts are everywhere: "No Church Tyranny" and "Remember the Plague, Remember HQV", but also a new, gentler one: "Give progress a chance." They are charter-supporters, obviously, but probably moderates, not the CDL. Nevertheless, it is not safe. She no longer has a security detail. She never felt comfortable with it, he knows. And it did make her more conspicuous. But coming here alone? That's reckless. Does she really think no one will recognize her? He finally sees her up ahead. She has a scarf tied over her head and is moving along quickly. It is the scarf he recognizes. She used to use it for family outings, back when they were still a normal family. The disguise may still work for everyone else. In public appearances, Marie Delacroix is always bareheaded. The neat cap of dark brown hair with its bold tracks of gray is part of her image. Reluctantly, and at a distance, he follows her across the square.

She seems to be moving toward a separate collection of people. These days, any demonstration creates an instant counter-demonstration. The videos go out and people simply leave what they are doing and join in. The crowds also feel more volatile, despite the steep increase in house arrests. Or maybe because of it, he thinks. The seasoned organizers would know how to control things.

© Springer Nature Switzerland AG 2020
P. Rørth, *The Unedited*, Science and Fiction, https://doi.org/10.1007/978-3-030-34624-9_10

"No, no, no," he says under his breath, "not that way!" She definitely should not be anywhere near these people. They are gathering in the forecourt to the old cathedral. He moves closer. The cathedral used to be a passive part of the scenery, a historical backdrop to modern life. Recently, it has come to life again. The raised, flagstone-covered area immediately in front of its massive doors is perfect for an impromptu stage. A group of purple robes have assembled there in a tight circle, triggering a rush of onlookers to the area right below. What are they up to? he wonders. Staging a public prayer? He realizes he can no longer see the scarf. The thought that she may have lost her disguise and will be recognized in the midst of this crowd sends a shiver of fear through him. He abandons his ambivalent distance and heads straight for where he last spotted the scarf, squeezing around people or rudely pushing them aside.

The scarf has shifted backwards and most of her head is now bare. He wants to shout and warn her, but he stops himself. She is close to the stage-like area, immediately below the purple robes. They are chanting something and surround a pale figure low to the ground. The person lying on the wet flagstones is, shockingly for this time of year, almost naked. His mother is staring at this poor creature. She starts yelling something. He wills her to be quiet and not attract attention. He pushes ahead and as he does, looks again at the near-naked figure. He catches a glimpse of the face. It is his brother, François, skinnier than he has ever seen him and with an expression equal parts pain, ecstasy and pure lunacy. The surrounding devotees are wielding thin sticks or whips. No wonder his mother is yelling. She seems to be trying to climb up on the stage, but loses foothold. When he reaches her, he is momentarily paralyzed. He cannot stop staring at the prostrate François. Marie finally notices him and grabs his arm.

"Help me up," she orders. "I need to get hold of him." Raphael complies and jumps up after her. Marie starts yelling at everyone, flaying her arms about. Her fury clears some space around François and stops the chanting and whipping. She drops down and takes hold of François' head. She touches his face gently. He whimpers softly. She pulls off her coat and throws it over him. "Pick him up," she says over her shoulder. "I have the pod close by." When Raphael picks up his elder brother, he is shocked both by the lightness of his body and by the haunted and uncomprehending expression on his face. François shows no indication that he recognizes him. Raphael clears his mind of anything other than two simple goals: follow his mother and keep a good grip on his brother.

Marie's natural authority helps them get away quickly. People simply move aside as she charges ahead. They appear not to recognize the former minister, however, only the determination of an angry mother. She takes the first street leading away from the square, then turns down another. Raphael follows. He sees a tall pod-park ahead. In the elevator, Marie inspects François' face and the visible parts of his body, shaking her head and mumbling something all the while. Raphael does not speak until they are safely inside the pod with François bundled into multiple thick blankets in the back seat.

"How did you know he would be there?"

"I had information," is all she says, giving him a quick glance. After clearing the pod-park, she sets the steering to automatic. She leans back in her seat and breathes out. "I've had his communications monitored for while," she continues. "I'm not proud of it, but I felt I had to. I still have friends with surveillance access."

"Who were those people? Why was he with them?" Even as he asks, Raphael knows the answer. For months, his parents have been going on about François being unusually distant, not eating anything and in other ways showing signs of religious mania. Raphael has been ignoring it. Actively. He has avoided visits home and has not answered messages. He only went today because the text did not mention François. To cover his shame, Raphael continues without waiting for an answer. "They are the purple robes, I understand that. But why the punishment? What is that supposed to accomplish?"

"Atonement for the sin of being a hyper," Marie says drily. "They believe God will reverse the block if they..." She sighs and shakes her head.

"But the block has nothing to do with hypers," Raphael says with exasperation. "It affects everyone."

"You know that. I know that. But no one is listening to sense any more. People want to find something or someone to blame. Getting rid of us is old news—plus it didn't fix anything."

"But François should know better. That's what I don't get. He's not stupid. He can understand..."

"It's not logical." She looks at Raphael steadily. "François hasn't been logical for quite some time. Haven't you noticed?"

Raphael looks out the window. He knows this—or sort of knows this. "But his work..." he says. "I saw he had another publication recently. He-"

"When he escapes into the abstraction of equations, he can be almost normal," Marie interrupts him, impatiently. "That's partly what misled us."

"And now?"

"Now I'm taking him home. And I'm not letting him out of my sight." She hears a fresh whimper from the back seat and turns around rapidly. François' eyes are closed but he must have twisted around in his sleep. The blankets have slipped off the skeletal body. "Just look at him!" she says with a sob.

Raphael looks. The recent marks on his brother's body are vividly red and the most convincing signs of life on the ghostly pale background. He turns back around quickly. "And then what?" he asks his mother, somewhat aggressively.

"What the fuck do I know?" Marie yells. She looks at her younger son with fury. Then she turns her head and stares straight ahead. They ride for a while in silence.

"I'm sorry, Rafi," she says softly. "I'm sorry I yelled at you. None of this is your fault." He glances at her and sees that she looks as remorseful as he feels. She touches his arm gently. "Thank you for coming today," she adds.

"Are you sure we shouldn't be taking him to a hospital?"

"I'd rather…" She takes a deep breath. "We've hired a nurse for home. She's brought IV equipment and everything."

"That's probably best, then." Raphael nods. A few moments later, he frowns. "I thought the purple robes were all about peace and healing," he says. "I thought they were harmless, essentially. Not like the grabbers."

"As you know, the purple robes are very religious. Harmless or harmful is a matter of perspective. But yes, like other unifiers, they want peace. Enforcers, or grabbers as you call them, want war. All they have in common is anger at the former government."

"Yet another reason why you shouldn't have been there."

"I had to go." She sighs. "Your father means well. He loves both of you very much. But the truth is, he can't handle this. These past months, he's…" She shakes her head. "He might just convince himself that François is an adult and therefore has the right to do whatever he wants. I couldn't take the risk."

"I know. I mean…"

"You should talk to him, Rafi," she says, very softly. "Talk to your father." She sounds sad. "He misses you."

Raphael finds no answer.

A little later, Raphael speaks again. His voice is dry and factual. "There's something else that unifiers and grabbers agree on: that mixing the populations is the way to get past the block. They just differ on the means to get there."

"That's true," Marie responds after a short delay. Her voice is back to normal. "I have to admit that this still makes me very uncomfortable. It goes against everything we've been brought up to think." She pauses. "Unless HQV has actually been eradicated, as they claim. We have no way of knowing."

"No way of knowing…"

"Not the most convincing of statements." She raises her eyebrows. "I know. But maybe they are right. Maybe it *is* gone." She stops, contemplating something. "For somatic technologies, I think a compromise is possible. We don't want to end up back in the dark ages." She turns to him. "It's not like anyone has a better solution, is it?"

"People are working on it."

"I know. But we can't simply sit back and wait ten years for a technological solution that may or may not materialize. Our society will split at the seams before then. That's a fact."

"Facts in politics? You're kidding me!" Raphael smiles. Marie smiles back. Then they remember their back seat cargo. The rest of the trip is silent. But it is a good silence.

When they get back to the family home, their actions are wordless, but well coordinated. Raphael opens the back and picks up the semi-conscious François. Marie grabs blankets and locks the pod. The front door opens and Anton steps out, looking mortified. Marie walks up to him and gives him a long hug. When Raphael reaches them, they each put an arm around his shoulders and guide the brothers inside.

*    *    *

"Rafi!" Eiko yells and starts waving energetically. "Rafi, wait up!"

He turns around and sees her running across the lawn. "Don't run," he yells back. "The grass is wet." She slows down and they meet in front of the massive biotech building. There is no one else around, neither protestors nor students. It is Saturday morning and a slow drizzle is falling.

"I haven't seen you for *ages*," Eiko says and opens her arm wide.

He bends slightly for the hug. "Busy... You know," he says and cocks his head at the building behind them. "And you?"

"The same."

"Saving humanity?"

"Don't tease." She slaps the arm of his heavy jacket. "We're doing what we can."

"I know." He looks somber. "I hope it..." He doesn't finish the sentence. Clichés have started to seriously bother him.

They look at each other, momentarily lost for words. Then they both start talking, at exactly the same time. "Have you seen Celia recently?" they say. They laugh.

"We are so pathetic, aren't we?" Raphael adds. "I haven't, though—seen her." He sighs. "We're having a bit of a break."

"Is it serious?" Eiko looks concerned.

"Might be. I'm not sure. I said some not-very-nice things."

"Hmm," Eiko says. "I'm kind of worried about her. She's gotten herself—or her company—involved with some very powerful people. I think it might be dangerous."

"In what way?"

"She says she can't talk about it." Eiko frowns. "It has to do with defense strategy or something like that—shields, for sure."

"Huang Shields?"

"So you *know* about this?"

"A little. Well, no. I'm guessing." He shrugs. "Have you seen them? Huang senior and Huang junior?"

"We've all met Leo." She pauses. "I don't think I've ever seen his father."

"Anyway, she's pretty good at taking care of herself, isn't she?" He makes a dismissive gesture. "Our Celia."

Eiko does not like his tone, but she has learned not to get involved in their skirmishes. "Let's get something hot to drink," she says and takes a step toward the building. "I'm freezing." She looks down. "And my shoes are soaking wet."

The giant glass doors respond silently to their IDs.

"So, how *is* the work going?" Raphael asks once they are seated with her lemon-ginger infusion and his black coffee. The canteen is practically empty. "Or am I not allowed to ask?"

"You are absolutely allowed." She bends over, pulls a pair of cotton slippers out of her bag and puts them on. The wet shoes and socks have already been pushed aside. "We are big on transparency these days," she adds as she resurfaces, making a face. "Don't get me wrong. It's a good thing, but... Well, no breakthrough yet. As far as I know." She shrugs. "I'm just a lowly student working away in my little corner."

"Is that the same corner as your father?"

"No, we're in different divisions. His is focused on identifying the intersecting agent."

"So, whatever is triggering the block right now?"

"Exactly. Some of the more obvious things have been ruled out. It's probably *not* another virus."

"That's good," he says. Eiko looks skeptical. "Or not?" he adds.

"Positive ID would be preferred, actually."

"I suppose." He tilts his head. "Will they find it, do you think?"

"I hope so. Because there's another problem. Maybe." She hesitates. "I probably shouldn't be telling you this... We don't know for sure it *is* a problem. It's just..."

"Oh, come on, Eiko." Raphael frowns, but with a smile.

"I haven't seen the data myself. My PhD advisor just told me—yesterday. Amita Singh. You know her, don't you?" Raphael nods. "She heard about it from some of her colleagues. It's relevant for our work, so..."

"So?"

"Well, it's a negative result—and not conclusive, but..."

"What?"

"They tried reverse-editing the HQV resistance alleles in blocked pre-em cells." She studies him. "To see if it was possible to rescue them."

"And?"

"Well, there's the question of what assay to use. They couldn't just go to the clinic."

"I suppose not."

"So the pre-ems were implanted into uteri of humanized pigs."

"That works?"

"It can—for early stages of development. They won't go to term." She looks slightly uncomfortable.

"Wait a minute—didn't the treaty with New Eden make all that illegal?"

"That damned treaty…. How are we supposed to make any progress when we can't even-" She stops abruptly and looks at Raphael. "Sorry. It's just so frustrating. Politics…"

"Makes the world stop going around."

"Anyway, the treaty. So, twinning is strictly prohibited." She pauses. "In fact, any human IVF procedure is out. It's illegal to manipulate unedited crossovers and natural hybrids in any way, even somatic. Abortions are out-lawed as well, although that's hardly relevant these days. The pig procedure wasn't mention in the treaty. I think they simply didn't know about it."

"Convenient." He raises his eyebrows.

"Yes. *If* they knew about it, it would probably be banned as well. Anyway, it didn't work."

"It didn't take?"

"It was only tried once, but no, the reverse edits didn't do the trick."

"So maybe the HQV edits have nothing to do with all this!" Raphael looks excited. "That would be huge."

"Or it's a problem of timing—or of the precise sequence used in the reverse-edit. It's just one negative result."

"But if the block has nothing to do with the HQV edits…"

"Then we understand even less than we thought we did." Eiko throws up her hands. "They *have* to matter," she continues. "We have lots of data on border hybrids and hidden uneds now. Everything points in the same direction."

"Hidden uneds?"

"Like Ben, remember? He lived here but he was—is—unedited. Somehow his parents and their family doctor got around the system. It turns out he's not the only one. Many of them are linked to…" She frowns. "Well, never mind. Some didn't know about their status, like Ben. Those that did know have mostly stepped forward voluntarily. As have the border hybrids. There are plenty of those. They used to be outcasts. Now they're welcomed with open arms."

"In domes?"

"They are in domes for everyone's safety, including their own."

"You believe that?"

"I do. Did you see how people reacted to the first live feeds? Of the babies, I mean? Imagine if they lived in a regular home. It would only take one determined psycho to-"

"Still—to be on display like that…"

"I've been told most of them are not. And the ones who are get paid handsomely."

"Well—whatever—but you were saying—they're all fertile?"

"They are, despite sharing all possible environmental factors with us. So it has to be genetic—the common edits. The lack of rescue is a negative result. Timing, sequence, pre-ems used, assay… Anything could be off. Maybe it has to do with the intersecting agent and how or when *that* works." She sighs. "We still don't know what it is—only what it's not."

"Despite all the time and effort—a whole division?"

"More. It's a huge effort. There are just so many possibilities."

"But you're doing something different?"

"Yes. I'm looking at genome structure and gene expression in pre-em cells and preceding stages, so germ cells—both oocytes and sperm—and early zygotes." She cocks her head. "Do you remember the hack that crippled the Reserve a couple of years ago?"

"How could I forget? My mother is still kicking herself for letting the government try to hide behind it."

"Yes, yes, I know that," Eiko says with a hint of impatience. "But there was an unexpected benefit. A lot of older, unclaimed back-up samples were found when they did the manual re-assignment: pre-em cells, eggs and sperm that are either pre-edit or early edit. The eggs aren't as stable as pre-ems, so they aren't functional any more. But most of the material is useful for chromatin and RNA analyses."

"And what do you compare it to?"

"More recent samples, mostly backups, that are donated to us." She pauses. "People are being extremely helpful." He nods. She continues. "I'm looking for changes around the HQV resistance loci—and elsewhere—that might be informative."

"Sounds reasonable."

"It's tricky, though. I've found some differences that seem to be real, but they are in low-abundance transcripts and mostly quantitative—not black and white."

"Complicated."

"Yes. Plenty of statistics required, that for sure. I have to deal with sample-to-sample variation *and* correct for sample degradation *and* not waste any of the material…" She raises her eyebrows. "And somehow interpret it all."

"You'll crack it."

"Someone believes in my project. Yeah!" She makes a fist and pumps it once with enthusiasm. "Thanks, Rafi." She smiles. He smiles back. "And you? How is your project going? You told me a bit about it a while back. Pretty peas, wasn't it?"

"Now you're teasing."

"Never." She adds another big smile. "What was good enough for Mendel is good enough for my friend."

"Not pretty peas—*interesting* peas… And it's going quite well." He jumps up from the table. "Let me show you," he exclaims. "The growth rooms are right over there." He points out the window at an adjacent building. "Come see my creative contributions to Biotech."

"I'd love to." Eiko tilts her head. "*Interesting* peas. I'm *very* curious."

A few minutes later, they are walking down a corridor in the plant biology building. Raphael opens a door with his wrist ID and indicates for Eiko to enter first. She does. She finds herself in a very large, very brightly lit room with plants everywhere. Except for narrow walkways, all of the floor space is taken up by low metal racks holding plastic pots and large planters. Some of the planters have green seedlings emerging in neat rows, some have only bare soil, but most are full of plants in various stages of development. There appear to be no seasons in this room. Eiko walks along a row of planters and notices the labels and support sticks attached to the plants. She sees tools and plastic trays everywhere, and the occasional fine mesh suspended from the ceiling. Overall, the room is slightly disappointing. There is just as much brown as green and not much of any other color. She had hoped for vibrant flowers and tropical lushness. She does not tell Raphael this.

Raphael leads Eiko to a corner section that is relatively cheerful. One row of his pea plants is in full flower. She stops to admire them but he nudges her on to the next row, where the pods are full to bursting.

"Open one." He points to a pod. "This one."

She picks the pod and gently pushes it open. Inside are six tightly nestled peas. To her surprise and delight, they are striped—fine white and green stripes running all the way around each pea. "They are wonderful," she exclaims.

Raphael smiles. "Now that one," he says, pointing to a pod on a different plant. It looks a bit darker from the outside.

She opens it and finds wine-red peas. "Wow," she says, sliding them out.

"Try one," he says. "They are all edible."

She does. It tastes different from a standard pea, a bit sharper, but very nice. "Yummy," she says. "What else?" she asks, in a slightly teasing tone.

"This one." He hands her another pod. Inside, she finds red-and-white peas, the stripes broader and less regular than the green-and-white ones.

"They are absolutely amazing," she says.

"Modifying the pigmentation was the easy part. Modifying the cell division patterns to get proper stripes—and coordinating gene expression with it—that was trickier." He sounds pleased, understandably so.

"Wow," she says, once again, rolling the peas back and forth in her hand. "And are they," she looks up, "useful in some other way?"

"You mean like producing antibiotics or extra vitamins? Nah… They're meant to be appealing. And tasty."

"That they are." She smiles. "So this is all—just for fun?"

"Not *just* for fun. We've learned a lot about regulation of cell division and geometric constraints in embryonic development. Also," he holds up a forefinger, "peas are an important protein source in a modern, balanced diet." He makes a face at this stodgy statement. She laughs. "Which means," he continues, "that improving their visual and gustatory appeal enhances our quality of life. A little bit."

"I get that," Eiko says, rolling the pretty peas from one hand to the other. "Actually, I do." She eats a few more, paying attention to the taste of each one. As she does, her eyes fall on the next row over. "Are those yours as well?"

"Yup. They're basically the next generation from the ones you're munching." He inspects the first two plants, feels their flat pods and starts to frown. He checks the labels and moves on to the next two plants in the row. "How very strange," he says, fingering more flat pods. "They should be mature by now." He points further down the row. "Look: the other ones are," he says. Eiko looks and nods. Raphael picks up a scalpel lying close by and uses it to open one of the flat pods. He looks puzzled. Eiko pulls his arm down and leans in to get a better view. Tiny, rudimentary peas sit along the seam. "The embryos—the peas, I mean—are there but they aren't developing," he says and shakes his head. "So the plants are effectively sterile." He checks the labels one more time. "How weird. They are genetically identical to those back there." He points to the plants from which they have just sampled the striped peas.

"Could it be chromosomal abnormalities?" Eiko asks. "Something segregating inappropriately in the offspring?"

"I don't see how. They are purebred lines and all I did was to put in a couple of transgenes, which are now homozygous."

"So how can these plants be sterile, if they are genetically identical to the fertile ones we just looked at?" Eiko keeps her eyes on the misbehaving plants in front of them. "It makes no sense." Her brow furrows.

"I think I know what it might be," Raphael says after thinking for a while. "I've heard of this phenomenon for other plants. I've just never come across it myself."

"What?"

"Second generation epigenetic effects. It's when a genetic change takes an extra generation to cause a phenotypic change. The environment is identical." He indicates the room. "You print out the two genotypes," he points at the two rows of plants "every basepair of it, and they really *are* identical. But it's a genetic effect, nevertheless. Just with a delay."

"Is it known how it works?"

"Most likely residual chromatin marks. Most chromatin marks are erased in the germline. But not all."

"So these are protein-based effects? Histone modifications and stuff like that?"

"Could be." He shrugs. "But they can involve non-coding RNAs as well, coating a locus or a chromosomal region. You know…"

"Yeah, I know. They can be tricky to analyze. I've got some really low-abundance RNAs in the regions where-"

"Holy shit, Eiko!" Raphael exclaims, grabbing her arm in excitement. "Second generation epigenetic effects!"

With a moments delay, Eiko's eyes open wide. She starts nodding. "Yes, yes, yes. That could be it," she says, while Raphael continues talking rapidly.

"Maybe there *is* no intersecting agent," he says. Second generation epigenetics could be… That's why the block is happening now and not when…"

"Let's think, let's think." Eiko furrows her brow and shakes her clenched fist. "The first generation homozygous edits were fine—healthy and fertile. Their children, mostly in their twenties and thirties, are healthy but *not* fertile.

We're essentially a population of grandchildless." She looks at Raphael, who is beaming like crazy.

"No, it doesn't fit," she says a moment later, her shoulders dropping. "Remember the original editees?"

"Sure. There were five of them, all homozygous for both HQV-resistance loci. One was killed during the crisis."

She takes a deep breath. "The remaining four are all grandparents now."

They both sigh, deflated.

"But..." Eiko looks up quickly. "Rafi, listen! They weren't selfers, like your plants are. I mean, their partners must have had unedited germ lines. The zygotes wouldn't have been homozygous until after the pre-em stage, whereas in the *next* generation-"

"Holy shit, Eiko!" Raphael repeats. They look at each other, unmoving, their eyes shining brightly. "Oh, man," he finally says. Eiko starts jumping up and down.

"So," she says a moment later, full of resolve. "How do we test this hypothesis?" She looks around and sees a nearby stool. Letting go of the peas in her hand, squished into a colorful mush in the excitement, she pulls it over and sits down. Raphael does the same. "There should be data out there to back it up—epidemiological data. We just have to look. Generations aren't separate strata over time. Well, normally, they aren't. The crisis did sort of reset everything."

"It did, but synchrony is lost relatively quickly. Those who got started early, in both generations, they should have arrived at the block longer ago. And..."

"Detailed family data with birthdates, pregnancies and failed IVF attempts must be somewhere at the Reserve. It's unlikely to be public, though. We'll need someone with access." She tilts her head, thinking. "Amita. Let's ask her. She knows people everywhere."

Raphael is following his own line of thought. "At the other end of the spectrum we have-" he says, then stops and looks pointedly at Eiko.

"What?" she says when she notices.

"Youngtwins. There are people in their twenties and thirties who genetically—and epigenetically—belong to the former generation. People like our pal Leo. He must be at least forty years younger than his oldtwin. I know there aren't a lot of them, but youngtwins like him are the perfect controls."

"In principle, yes." Eiko makes a face. "Except youngtwins are specifically not allowed to have children according to the treaty. I never understood that part."

"Maybe those crafty priests from New Eden somehow know that youngtwins can help us understand the block."

"Nah... That's too paranoid."

He nods. Then a mischievous smile appears. "We *could* always make a world full of youngtwins," he says, eyebrows raised.

"Don't even *try* going there." Eiko gives him a hard look. She *has* thought of this option before. She has even wondered why it has not been part of the public debate. Once the cleanup from the hack had been completed, it was feasible. There is even a decent pool of unborn uncles and aunties that could get a chance. But she will not be the one suggesting this.

"Sorry. I wasn't serious." Raphael adds a quick smile. "But I suppose the priests might at least have considered that possibility."

"They might have." She nods. "Anyway, back to hypothesis testing: I'm useless as a youngtwin. My oldtwin was my sister, so I'm the same generation as everyone else my age." She laughs. "Peculiar statement, isn't it?"

"I get what you're saying, though." He frowns. "Somehow I don't think we should ask Leo for help. He seems a bit... unpredictable."

"I agree. And we both know there's something funny going on with Huang Shields. I don't want to get Celia in trouble."

"So, what are our options? We have this perfect control group, but we can't ask them for help? Not even when the future of humanity is at stake?" He raises his hands, palms up, to accompany his grandiose statement.

"Plenty of youngtwins are nice people, as concerned about the future as everyone else," Eiko says. "I've gotten to know quite a few through my young-twin support group." Raphael looks surprised. She shrugs. "It's all very low key and non-political. We just talk, compare stories and commiserate. And eat cake." She smiles. "I could ask some of them—discreetly. Accidents do happen... I'll figure out a way." She stands up. "But first, let's run your idea by Amita."

"*My* idea?"

"I think that's fair. We've been staring at this problem for over two years and no one has said anything about second generation epigenetic effects. Including me."

"OK, I accept." He adds a brief nod and a stern expression. Then he frowns, serious for real. "Are you sure it's a good idea to talk to your PhD advisor about this?"

"Absolutely—but not just because she's my advisor. She's really smart, so maybe she'll think of some angles we haven't covered. She's also very well connected at the Reserve *and* she really wants to solve this damned thing." Eiko tilts her head. "In this, her personal and professional desires overlap one hundred percent."

"Maybe we should consider one more thing first," Raphael says, staying seated. Eiko sits down again. "Possible solutions. Does this hypothesis about the block's genesis—assuming it is correct—get us any closer to an actual solution? If it doesn't, we might not want to… You know." He gestures openhandedly. She nods. He continues. "I'm also thinking about what you told me earlier—the lack of rescue in reverse edited pre-em cells."

"I think it *does* get us closer. It forces us to focus directly on the two resistance loci and the precise changes that were made in the common edits. I'm not sure the reverse edits were completely accurate. The coding changes were reversed, yes, but no one checked other RNAs in the region—non-coding and rare ones, for example—that may have been perturbed."

"Is that the kind of thing you've detected?"

"Some candidates, yes. They appear to be altered in historical samples versus current ones. We can combine this information with chromatin analysis—local and further away—to see what changes show up and when."

"But simply reversing it—that's not a solution anyway, is it?"

"Not a good solution, no."

"Ideally, we'd want to keep the HQV resistance effects but restore whatever else has been perturbed."

"Exactly."

"It *could* be the coding changes that cause the block."

"I very much doubt it. But we'll have to test all the options."

"Test how? You said the somatic rescue didn't work in the transplant assay—the humanized pig thing."

"That's to be expected if the block is due to the kind of effect you're proposing. DNA changes in pre-em cells would probably be too late. We'd have to trick the system into a restart of proper regulation. If it's an RNA that's needed, maybe we could add it directly—transient rescue—and let the system recover from the genomic locus during next germ line development."

He nods. "That sounds… reasonable."

"Plus we have the youngtwin angle. They might be willing to give us some samples to work with. Positive controls."

"Would that be legal?"

"I'm not sure. Better not to ask."

"Right." He smiles. "We're just a couple of naïve PhD students, who don't know anything about political treaties." That will hardly convince anyone, he thinks to himself. "I didn't know any better, you honor." What a great defense. Particularly for him, with his mother being who she is and having had that particular political fiasco… But surely, there's a bigger picture here?

"Does that make you feel better?" Eiko says, bringing him back around.

"It does. Let's go." Raphael slips off his stool. "Maybe your advisor has heard all this before and there's a perfectly good reason why no one is pursuing it."

"I doubt it."

Before they leave, Eiko asks if she can have a few more of his special peas. Raphael lets her pick whatever she wants. Meanwhile he is standing guard, looking fondly across his rows of very interesting pea plants.

\* \* \*

"I think it's a *brilliant* hypothesis," Amita says. "How extraordinary!" It is an hour later and Eiko and Raphael are sitting in her office. They have just finished telling her about the "second generation epigenetics" hypothesis and some of their plans to test it. "How on earth did you come up with it?" she asks.

Eiko indicates to Raphael that he should explain. He does.

Amita keeps shaking her head and smiling while Raphael is talking. "I have *no* idea why no one else has thought of this," she finally says. Then she frowns and moves her thick, black braid from her back to over her right shoulder. She slides a finger slowly over each twist. Eiko recognizes the movement. She thinks it means Amita *does* have an idea but is unsure whether to share it. Amita clears her throat and speaks again. "Well, maybe I do. It is probably due to our over-fixation on the 'intersecting agent' hypothesis. Everyone seemed to jump on that bandwagon at once."

"It *did* make sense," Eiko says, slightly stung on her father's behalf.

"Of course it did," Amita says. "That's why everyone jumped. It still does. Making sense doesn't necessarily make it true, however."

"Does that matter? At this point?" Raphael asks

"It shouldn't. But people can be surprisingly stubborn when the alternative is acknowledging that they may have been wrong. And given that supporting evidence will mostly be hard to come by-" Amita senses that Eiko and Raphael are about to interrupt. She stops them with a gesture. "I can guess where some very useful information might come from. Youngtwins are a special case. But most approaches would not be legal." She looks stern. "So I do not want to hear about it." They both nod. "And we may not be able to use such data officially." More nods. "I'll talk to my friend Sanjay and see if he can make a compelling case from the epidemiological data. That is something we *are* allowed to look at and it may carry enough weight." She looks from Raphael to Eiko and back again. "I'm really impressed, you two. Good thinking. No— brilliant." She smiles.

"If we *do* find the exact cause," Eiko says, pensively, "I mean, if we identify the molecular change that causes pre-em cells to be blocked *and* find out how to selectively revert it, one *could* argue that it would work whether or not there is an intersecting agent. So there is no need for anyone to feel threatened." Amita looks at her with amazement, Raphael with a knowing smile.

"So, we'll have what my mother would call a cake solution," Raphael says. "A solution that is expected to work no matter which view of reality—yours or your political opponent's—is correct. Everyone will feel that they have been vindicated—and happily eat cake."

"Politics," Eiko says, "yuck!"

"You are right, though," Amita says to Eiko. "Absolutely right. This also means that the practical aspect of your project remains largely unchanged. You just have that extra bit of motivation to keep you going."

"A bit?"

"A mountain-sized bit." Amita smiles at both of them. "And there will be plenty of support in the background. I promise you that."

Eiko looks at Raphael and silently mouths: "You see?" He nods.

They leave Amita's office with a plan to meet again in a week. Sanjay can do the required analysis of the medical records in a few days. Amita assures them he will be motivated.

When they get back down to the canteen, it is pleasantly busy. They decide that they deserve a good lunch.

At the salad bar, Eiko leans over and whispers in Raphael's ear. "Are you sure it wasn't you who got the genius edits?"

"I'm sure." He smiles as he says this but then quickly looks away. He cannot bear to talk about François. Not today.

"Well—you're doing pretty good for a regular guy," Eiko adds.

As they turn toward the main part of the room, two sets of arms fly up and wave them over. The arms belong to Jun and Martha, both PhD students from their year. Lunch passes noisily, with plenty of talk about useless PhD advisors, long hours and experiments proving to be impossible, followed by jokes about the easy old days of pre-adulthood. It seems so long ago. Eiko and Raphael both understand that their recent conversations are not to be shared, not yet. They exchange occasional complicit looks, which are conveniently misunderstood by their fellow students. In retrospect, Jun and Martha agree as they take back their trays, they always knew Eiko and Raphael would become a couple some day.

\* \* \*

"Standing out here is really fucking stupid!" he says to himself, jumping from one foot to the other in order to stay warm. "But at least you deserve it."

A middle-aged, well-dressed couple walks right past him. As the oversized doors to the hotel open for them, the man turns around and looks disapprovingly at Raphael. Judging him to be the kind of stiff who would not hesitate to complain about a crazed loiterer to the hotel staff, Raphael refrains from engaging in a staring contest. Instead, he turns to move off. Soon after, he returns to his post. He sees the couple, minus overcoats, as they mingle with the rest of the moneyed people inside the hotel. The view from the sidewalk is limited by extensive frosting on the windows, but it is still pretty good: A slice of the Grand Hall plus most of the stage. The stage is currently empty. Raphael continues his game of attaching a most likely profession to each person that passes into his narrow view. Half of them look like bankers, he thinks. This one must be an arms-producer: The way he struts, chest forward, and the way he holds his head while talking. He wants to be seen and heard. The finer nuances of information, resources and intentions around him are secondary. Whereas this one—Raphael judges the outfit, the body language, the amount of listening versus speaking, the discreet confidence—she must be a top-level lawyer or analyst. The arrogance of these people, he thinks. And they are all useless—or worse. Once again, he tells himself to leave. Now! But then the elegantly dressed bodies rearrange themselves and turn toward the stage. Something is happening. Finally. He decides to stay put.

The introduction seems long, the prize presentation itself much shorter. Celia comes on to the stage from the right. She looks absolutely stunning. He should not be surprised, but he is. He stares. Her dress is long, a warm, dark green color, and shiny. She is wearing serious jewelry, which he has never seen her do before. It sparkles in the stage lights. Even her hair looks different as it tumbles generously over one shoulder. She shakes hands with the over-smiling presenter, who has now become a bland, forgettable man, and receives the trophy with both hands. The presenter steps out of the limelight and Celia turns to address the audience. She must be saying something humorous. The big smiles are out and shoulders shake with indulgent laughter. Their attention is absolute, undiluted. He looks at her as she scans the proximal audience for questions. He wishes for her to raise her eyes and look at him, instead. He wishes he were inside, riding the elevator, waiting to meet up with her in one of their arranged coincidences. They would spend the night together—silly fun with the minibar and everything that follows. But tonight he was not invited. He had not expected to be, he reminds himself. Nevertheless… Now someone is joining Celia on the stage. An older man, short, slender, with silver hair. He puts a hand on Celia's arm. She does not brush it off. He indicates the trophy and adds a few words to the audience. Raphael recognizes him. Victor Huang. Of course it is. He backs away from the window, quickly. "You idiot," he tells himself as he starts to move away, "you bloody idiot." He does not see Celia glancing up and catching the sudden movement in the window. She knows his posture, his movements and his habits. She turns back to Victor and reanimates her professional face.

Raphael walks for hours, to keep warm and to think. His thoughts start with Celia and his stubborn infatuation, but they end up in a very different place. He and Eiko spoke with Amita again earlier in the day. Sanjay, Amita's statistician friend, was there as well. His findings from the database were consistent with Raphael's hypothesis. They were all very excited about it. Raphael's mind jumps ahead. He plays with ideas for experiments to do and tries to root out possible caveats. He follows each thread, step by step. Concentrating on this is not hard. It is fun. He even indulges himself in imagining a favorable outcome. He feels the deep satisfaction of having worked it out. He sees joyous reactions from his parents, friends, colleagues and even from strangers. A full smile comes to his face. After that, other thoughts crowd in. It was something Eiko said… or Amita… No, they saw it, the four of them, in a newsfeed. Someone got a text and Sanjay called it up on his screen. It was the minister of science—Gareth Svensson—giving a statement. He had just come from a visit to the domes and he had seen the future, he said. He seemed very emotional. The mistake of the past had been to rely too much on science and biotechnology, he continued. This had been their downfall. New attitudes were needed. He went on to say that the HQV virus was, fortunately, also a thing of the past. "Where did he get *that* from?" they asked each other in the little room. "And he calls himself minister of science?" They watched the rest of the feed, dumbfounded. Amita started talking, angrily, about ignorant deniers. Sanjay wondered which faction had won the minister over, and how. Raphael did not call home afterwards. He could not face a discussion with his mother—or talking to his brother. What would happen to their project? Would it be shut down? He did not dare ask what the others thought. They walked out of the office like zombies.

Forget that stupid politician, Raphael tells himself now. Think, instead. Focus and think. He finds himself close to where he started, in front of a familiar all-night café. He goes in, gets a black coffee and a slice of a heavy-looking chocolate cake. He finds a seat near the window and starts filling his notepad with ideas recovered from the walk.

At some point later, coffee and cake long gone, nighttime patrons come and gone as well, he feels someone close by. He has no idea how long she has been standing there. She is still wearing the green dress but now a long, black coat as well. The jewelry is gone and the hair tied back, but she still looks far too glamorous for the place.

"May I?" Celia asks, indicating the bench across from his. She has already started to remove her coat.

"Of course." He closes the notepad.

She slides in. It is her side, he suddenly remembers. It is their table. She places the folded coat on the bench. "Are you feeling a tad sentimental?" she asks.

"Maybe."

"Me too." She looks sad. "I thought I might find you here."

"I didn't know that I'd be here."

"That's why I'm better paid than you are." She makes a face. He returns it. They smile, briefly.

"So, I gather your work is going well?" he asks, slightly forced. "Your business?"

"Yes, very well." She straightens up. "Everyone wants to use our modeling approach these days—for all kinds of things. That's why it took me so long to get away from the gala."

"I saw you."

"I know."

A short silence.

"Where's the trophy?" he asks, unexpectedly. "And the…" he indicates her bare neck.

"Where it belongs."

Another short pause.

"And did you deserve it?"

"Of course I did." She raises an eyebrow. "The simulations created at Response Modes are unique—and pretty damn sophisticated."

"Your partner at Response Modes, where was she tonight? Wasn't the trophy for the company?"

"It was." Celia pauses. "She never goes out. She hates publicity and dislikes people."

"She's the genius behind the scenes?"

"You could say that." Celia smiles, mysteriously.

Raphael does not to pursue this. Instead, he asks "simulations of what, these days?"

"I can't tell you. Not yet."

"Of course not." The sarcasm is wrong—he can feel it. "I'm sorry," he adds, in a much softer tone. "I didn't mean to…" She shrugs. "Celia," he starts again, his voice heavy. She looks at him. He takes a deep breath. "I've been wanting to apologize."

"What for?" She seems genuinely surprised.

"For the things I said last time we were at your place."

"When was that? A month ago?"

"Two, at least."

"It's been a bit crazy." She sighs.

"So, I wanted to apologize."

"Yes?" She waits.

"For the things I said about you being a hyper and all that. It wasn't OK. I'm really sorry."

"We were arguing. Words fly. It's not a big deal."

"So you're not still upset?"

"About that? Shit, Rafi." She shakes her head. "Of course not. We live in a hard world. People say a lot worse."

"So that's not why you haven't…"

"No, no. It's just been a bit…. crazy. As I said." She forces a smile. "And how about you? Eiko said you might be doing some work together, but she wouldn't give me any details. She sounded very excited about it, though. And happy."

"It's good—It's…" His eyes are wide open. His mouth is eager to tell it all. But he stops. "I can't tell you. Not yet."

"Good one!" She smiles, poignantly. "I suppose I deserved that."

"No, it's not like that. It's just…"

They are both silent again. She looks at their reflections in the window. He looks at the empty coffee cup that he is twirling back and forth.

"Rafi," Celia finally says. "Rafi," she repeats and looks at his face until he returns the favor. "It might be impossible to reach me for a while. For a long while." Her voice is serious, flat and steady.

"What? Why?" His voice betrays both confusion and hurt.

"I may have to disappear." She rolls her eyes and puts on a stiff, fake smile. "I'm about to become a terribly toxic acquaintance." The joking tone and clowning expression don't quite work. She drops them. "Seriously, I can't have contact with anyone."

"But you just got this fancy award. And your work is-"

"It's complicated. People will say-"

"I don't care what they say!"

"You should care. Some of it will be true, I'm afraid. Don't try to find me, Rafi. And don't try to defend me. It's best you don't get involved."

"But Cee," he pleads, "I can't just…"

She reaches out a hand to touch his face, but draws it back again. After a couple of deep, slow breaths, she presses her lips tightly together lets her eyes go blank. "I've been sleeping with them," she says, off-hand but not at all casual. "The Huangs. Senior *and* Junior."

He cannot hide his shock. "What?"

Her face, her rigid expression, starts to slip, but she catches it and regains control. She picks up her coat and slides out of the bay.

"Take care of yourself, Rafi." She stands. "And if you have any shares in Huang Shields, sell them. Tonight." She starts toward the door.

"Shares?" he says, but does not move. "Me?"

He turns his head when he hears the jingle of the café door.

"Cee," he says, softly. "Come back."

# Part III

# 11

## Return

Eiko closes her eyes and turns her head to catch the warmth of the sun on her face. The calendar may say spring but the air still has bite. She hears the muffled sound of the shuttle arriving and opens her eyes. As she does, a solitary figure at the edge of her vision catches her eye. She is not sure why. Everyone boards the shuttle with her and the figure no longer stands out. The ride is only five minutes, so she remains standing, looking at the busy streets as they glide past. She feels a sudden fondness for the city and everyone in it, despite all that has happened the past few years. Maybe what she feels is hope.

Her mother had asked her to stay over, as she always does, but did not protest when Eiko said she would go back to her own place. She knows Eiko will be back in a couple of days to free her mind at the piano. Besides, Eiko's old room is now a guestroom—at Eiko's insistence. Her new place and its lack of distractions ought to be ideal for thesis writing. She knows it is time to get serious about that. She leaves the shuttle with her mind on the days and chapters ahead. She turns each corner automatically. When she reaches her apartment building she does not, however, use her wrist ID immediately. She stops abruptly and turns around.

"Why are you following me?"

"I…" He slides his hood back so she can see his face. "I'm sorry if I scared you."

"Ben!" She exclaims and wraps him in a spontaneous hug, a response they are both grateful for after the fact. "Oh, Ben, it's really you!" She pulls back and looks carefully at his face. Her first impression is that he has aged considerably. Maybe it is just the odd-looking beard—or the rough living showing through. She sees *and* smells that. A passing pedestrian turns his head to stare

© Springer Nature Switzerland AG 2020
P. Rørth, *The Unedited*, Science and Fiction, https://doi.org/10.1007/978-3-030-34624-9_11

at Ben, who quickly pulls the hood forward again. Eiko forces herself to hold back questions and instead turns around, swipes over the sensor and opens the front door.

"Come on in." She moves as if to take his hand but he keeps his arms firmly by his side. He does, however, follow her inside. "My place is on the third floor," she adds, choosing the steps over the elevator and looking over her shoulder repeatedly to make sure that he is following her. Only when they are inside her apartment does she continue talking. "I am *so* happy to see you, Ben. Alive and… We had no idea what happened to you." He lifts the hood off his head and she smiles at the red hair, still visible under the grime. "You probably want some water." She moves a few steps toward the kitchen area. "And a shower, perhaps? To warm you up."

"Yes, thank you. I think I need that," he finally says, and returns her smile.

His voice and his smile seem unchanged, which is a relief to her. She fills a glass with water and hands it to him. He drains it. The glass is filled and drained again. For the third glass she gets orange juice out of the fridge. "Oh, Ben," she says, touching his sleeve. He adds another "thank you" and a weaker smile. He sits on a chair while drinking the juice, his eyes slowly drooping. It's the indoor heat, she thinks, and he's probably dead tired. She hurries to her bedroom in search of clothes that might be appropriate for him, checks the shower room and starts the water running. She comes back with a fresh towel, a new toothbrush and a pair of loose but heavy pajamas.

"All yours." She indicates the shower as she gives him another reassuring smile.

"Thank you," he repeats, with an expression of relief. They will both get a bit of solitary time to adjust to the situation.

When he comes back out, he has put on the clean clothes. His hair is wet and his cheeks are flushed from the heat. She has put a plate of cookies and various mismatched items of food on the low table next to the sofa. A steaming cup of tea awaits him there as well. She is sitting across from the sofa with a cup of her own. "I thought you could do with more warmth and more sugar," she says. "Nothing too exciting to eat, I'm afraid…" She smiles, hesitantly. He sits down in the corner of the sofa and pulls up his bare and bruised, but clean, feet. She is happy to see that he starts on the cookies immediately.

"How did you find me?" As soon as she says this, she remembers the figure hovering at the edge of her vision. "Did you follow me from my parents place?" she adds, keeping the tone light.

"Your house was the only place I could think of going to. But once I got there, I wasn't sure what to do. I must have looked a fright. Your parents might not… So I just waited nearby to see if I could catch you."

"I go once or twice a week," she says, "to play—and to get a proper dinner. It's a good arrangement. I like having my own place." She gestures around her. "Even if it is tiny."

"It's nice."

"But tiny," she repeats, smiling.

"I heard you playing, I think, from outside the house. It sounded wonderful."

"Really? I'm working on the Chopin ballades these days. I love them but they are fairly difficult. Concentrating on them allows me to… When I have time, that is. I'm trying to get serious about writing up my thesis. I'm done collecting data, so I really should get going." She sighs. "There's just so much to check up on first. Procrastinating; I suppose that's what I'm really doing." She goes on talking about her thesis and the papers she needs to read, while asking herself why she is doing this when there is so much else to talk about. Glancing at Ben, she sees that it was a good choice. His eyes are drooping again. A couple of minutes later, they are fully closed and Ben's cup is tipping dangerously. She rescues the half-empty cup and finds an extra blanket. He slides further down the sofa and deep into sleep.

When Ben opens his eyes again he is only half awake. While he goes to use the bathroom, Eiko rearranges the sofa and gets him some more water. Before he drifts off to sleep again, she asks him the one question that cannot wait. "Is your ID still active? Should I expect the police or whoever to be showing up at the door?"

"No, it's…" He holds up his right arm to show her. The wrist-link has been ripped off and his skin is heavily scarred. "Gone. I guess someone wanted the link-up badly enough to risk my ID. It's pretty rough out there." He shakes his head slowly.

"Why didn't you…" She starts, but stops again. "Never mind. We'll talk about all that later. For now, just get some more sleep. You clearly need it." She frowns slightly. "I'll have to go out at some point," she continues. "If I'm not here when you wake up, just help yourself to anything you need—food or whatever. But don't go out, OK? Stay here. Unidentifieds can get picked up on the street at any time." She glances at the walls. "You don't need to worry about the house computer. I had it disabled as soon as I moved in. It seemed so strange, you know, talking to the walls. I didn't…" She looks over. He has fallen asleep again. She wonders briefly if she is doing the right thing here, all things considered. She decides not to overthink it. Ben needs a friend.

\*    \*    \*

"How long have I been asleep?"

"A day and a half—minus a few minutes here and there." Eiko opens the blinds. "It's morning again."

"Really?" He slides to sitting position. "Sorry about that. And sorry I just barged in on you, without warning."

"I figured you had a good reason."

"Hmm." He nods and accepts the coffee she hands him.

"Let me get you some breakfast. You must be starving." He smiles in response. "Then we'll talk," she continues. "I have all day."

Eiko gets busy in the kitchen area while Ben tidies up the sofa and himself. She brings over a laden tray and they both sit down. He starts eating while she sips at her coffee, glancing at him frequently.

"You cried out in your sleep a few times," she says.

"I did?" He looks up, a smudge of red jam in the corner of his mouth. "What did I say?" He puts down the piece of bread he was halfway through. He has been eating too fast, he realizes.

"A few times it was a like a shout—'No' and 'Get away'."

"From my travels, I suppose." Ben smiles enigmatically but does not expand on it.

Eiko frowns, remembering what his wrist looked like. "You also said a name. Lilly, perhaps?"

"Tilly, probably," he says with a sad smile. "She was my aunt, my mother's sister. She was... very special." He looks into the middle distance for a while. "She made me feel at home. Like I belonged. And then..."

"Then?"

"She died. She had cancer. It was horrible." He looks down at his coffee cup, tipping it back and forth.

"You met her over there—in New Eden?"

"Yes." He sighs. "I lived with her and the rest of my mother's family. I never even knew they existed."

"And you parents? Jack and Bella?"

"They died in that accident. I know that now. The rest was just... wishful thinking."

"That must have been-"

"It's fine," he says quickly. "I've gotten used to it."

"But your aunt... Tilly?"

"That still hurts, losing her."

"You had gotten close?"

"Yeah. Somehow we just got along really well. She'd always tell me how much I..." Happiness suddenly floods his face. It disappears just as quickly.

"She suffered a lot, when she was… toward the end." He looks directly at Eiko for a moment. "It rips everything out of you, you know? When you can't do anything to help, only… watch."

"Tell me more about her," Eiko says gently and tilts her head. "Tell me about your time with your family."

Ben starts talking. He describes the farm and the village, but mostly he focuses on Tilly: their time together in the kitchen, cooking, baking and exchanging stories, her amazing vegetable garden, her work at the village school. He mentions Harold and his children, but only briefly, and Eiko senses that she should not ask for more. Finally, he tells her about Father Len, their friendship, their surprisingly open chats and his church services filled with music. It reminded him of listening to her playing the piano, he says, and looks at her with another sad smile. She smiles back, cautiously. He mentions Tilly again, but the words seem to have run out.

"That's tough luck, cancer," Eiko says, when he has been quiet for a while. "Was she older than your mother?"

"No, almost the same age. Forty."

"But I thought…" She looks puzzled. "Oh!"

"They were quite different, though, growing up. My mother was the local beauty, apparently, and also quite the rebel. Headstrong, Tilly called it."

"Really?" Eiko looks skeptical. "Your mother seemed so…" She adds an apologetic smile. "Of course, I didn't really know her."

"I think life here didn't turn out quite as she had hoped. She wanted… well, it was complicated."

"I'm sure." She remembers once having lots of questions about his parents' situation—and his—but they appear to have lost their urgency. "And you father's family, did you find them?"

"No. Well… I never met them. They didn't want to. They decided a long time ago that they never had a son called Jacob."

"Jacob?"

"My father's real name. And, oh—this is kind of funny—Hatton is the name of the village. It was never anyone's last name."

"Confusing."

"Partly." He shrugs. "It got me there, I suppose."

"It sounds like a nice place," Eiko says, carefully. "And you did find your family. So, I guess it was worth it. Crossing over, I mean."

"I suppose it was. But then…" He looks away.

"Was that why you came back? Because you lost Tilly?"

"No, I mean, yes." He sighs. "It's part of the reason."

She looks at him for a while, but he does not volunteer anything further. "How *did* you get back, by the way?" she finally continues, trying for a somewhat lighter tone. "Supposedly, the border is even more tightly sealed than it used to be."

He perks up. "Believe it or not, some of the old network still works."

"What old network?"

"For crossing over. Aunt Vera used to be part of it, on this side."

"Oh, yes, Vera Weiss." Eiko remembers having heard the rumor.

"It turns out *her* family name was Birch, originally." He shakes his head, but smiling. "I met a real nephew of hers—and her older sister—well, Vera's adopted family, anyway—and this amazing old woman, Margot, who used to be her... mentor, I guess you could say."

"*They* helped you cross the border?" She sounds skeptical.

"Yes. I stayed at the Birch place for a while and learned the route. They still had contacts near the border. Father Marius knew about the more recent changes."

"Who's he?"

"He's the nephew, but also..." Ben tells the story of Tilly and Marius, slightly altered. He talks about Margot and the Birch family. When he starts to describe his trip back from New Eden, Eiko leans over and whispers urgently in his ear. "Sorry. I shouldn't have asked about that. I can't be absolutely sure the walls don't have ears."

"Right." Ben pulls back quickly. "Right. Well... It was a tough journey, that's for sure." He stops, looking worried and a bit confused. He pours more coffee, now lukewarm, and finishes what is left of the bread.

"Do you know what's been going on here?" Eiko asks.

"You mean the fertility problem?"

"Yes." She nods. "The block, people call it. An apt name, actually."

"Father Marius told me about it. He..." Ben stops and cocks his head. "So it's true? How bad is it?"

"It *is* true. And it's pretty bad—both the medical crisis and the political mess it has created." She shakes her head. "It's been general knowledge for close to four years now. People are starting to get desperate. Many have become obsessed with New Eden."

"People in New Eden don't know about the block. They just know that there is some sort of unrest over here and that the borders therefore need to be defended."

"Hmm. Well, some of them clearly *do* know—that Father Marius, for example, and those in power. Not surprisingly, they're trying their best to take advantage of the situation. They believe New Eden has a firm upper hand.

The truth is, everything is pretty volatile." She frowns. "Do you know the old government was ousted?"

"I heard."

"The dome scandal made that inevitable. Marie Delacroix—Rafi's mother, you know?" Ben nods. "She was smack in the middle of it. Anyway, it hasn't made much of a difference. The current government can't—or won't—do much, except blame the old government and try to avoid everything blowing up on their watch. They are internally split about how to move forward. Some still believe in technology and the Reserve, others think the answers lie with New Eden and places like that. Everything is still up in the air." She pours some cold coffee for herself and has a sip. "But it will come down—soon. Unfortunately, the government's position seems to be drifting in a bad direction. The new minister of science is an idiot." She shrugs. "That's just a fact. The minister for state security and defense is a key post. That used to be a CDL guy." She looks at Ben and sees his confusion. "Charter Defense League. They are the most vehement supporters of how we used to do things." Ben nods. She continues. "Now it's a woman from the other side of the divide. They've also become much stricter about security."

"Will there be a war?"

"With New Eden? I don't think so. The hawks had some serious setbacks, like the Shield scandal." She scoffs. "That put a bit of a muzzle on them."

"Huang Shields?"

"Yes, Huang Shields—those greedy bastards. Believe it or not, Celia was the one who…" She stops herself. "I'll tell you some other time. Or she'll tell you herself." She frowns. "Anyway, a relatively new party, the PUP, is gaining ground fast. I'm afraid they'll win big next time there's an election."

"Why afraid?"

"They creep me out. All this newfound humility… They are either too naïve or too damn cunning. I can't tell which. They think outcrossing and abandoning Reserve technologies altogether is the answer. In the short term, exchange and lots more hybrids, in the long term, an actual union with New Eden. All on *their* terms, of course."

"Hybrids?"

"Mixed parentage—edited/unedited." She looks at him for a moment without saying anything. "Have you heard about the giant domes?"

"I have." He looks at the empty cup in his hand.

She goes quiet again. "Why did you come to my place, Ben?" she finally asks, looking weary rather than truly puzzled. "I'm glad you're safe, obviously, but… Why didn't you go to a dome? Unedited, and with your background… They'd be happy to have you, you know."

"Do you want me to leave?" He puts the cup down.

"No, that's not what I meant. I just…" She looks straight at Ben. "Sorry, I don't know where that came from," she says, but keeps looking at him. Gradually, something shifts in her. "No, wrong, I *do* know where that came from," she adds, with an edge of bitterness and hurt. "Why did you leave like that, Ben? From the transit center." She takes a deep breath. "Why did you sneak away in middle of the night, without telling anyone?" Her voice has grown harder. "We were there for you. *I* was there for you."

"You couldn't go with me. Because… you know…"

"That doesn't make it any better, you idiot!" The anger bursts through so abruptly that she has to stand. The cups rattle. "Not a word. That was cruel! *And* selfish. I never thought you'd… You shocked me. You know that? I thought I knew you. I thought…" She takes another deep breath. "Well, never mind…" She sits back down. "It doesn't matter any more."

"I'm sorry." He sounds contrite. "I really am. It *was* selfish."

"No, I'm sorry. I shouldn't have… It's just…"

An uneasy truce settles.

"So," Ben starts, with obvious effort, "the mixed… the hybrids. Are they children?"

"And adults. As I said, they're still in isolation." She sighs. "But that will probably change soon. Domes are not a long-term solution. Everyone understands that."

"And does everyone believe that HQV has been eradicated?"

"HQV?" She looks surprised. "No, not everyone. Not me, my parents and… But many do. It's what the slick negotiators from New Eden keep saying and it's what people want to believe."

"I came here," he says and breathes deeply, "I came directly to you because I know HQV is still out there."

"It *is*?" She stares at him, with intense focus. "How do you know this?"

"Mostly from Father Marius." Ben leans backward, defensively. "He told me he'd had his suspicions for a while. There's something called possession. The Church claims it's an affliction of the mind, caused by evil spirits. But it's obviously contagious. Anyone who is deemed affected gets put away—quickly and forever. The farms or villages they are from are quarantined."

"So, it could be HQV." Eiko's eyes dart back and forth, eagerness fighting with fear. "It could also be some other infectious disease. It sounds like the cases are rare."

"Most people in New Eden hardly ever travel, and never very far. Some priests do, but they tend to limit bodily contact. Between that and harsh quarantine procedures the spread would be limited, wouldn't it?"

"I suppose so."

"The symptoms fit with what we learned in school about HQV infection."

"Is there any way to find out for sure? If we can show that HQV still exists, it would change everything here. *Everything*."

"Father Marius gave me some vials. He said they were blood samples from sanctuary patients. That's where they take them, by the way, to so-called sanctuaries. No one ever leaves a sanctuary, he told me. Some of them may survive, but they have to stay. They become-"

"Where are they?" Eiko interrupts him, impatiently. "The samples? We should-"

"I lost them."

"No!" She looks crestfallen.

"They were stolen along with my wrist-link." Ben sighs. "The thieves probably threw them away. Figured they were junk."

"Shit, shit, shit!" Eiko says, and lowers her head into her hands. She stays like this, hunched over in her chair, for a while. "But this Father Marius was convinced?" She asks, looking up again. "He was sure it was HQV?"

"He suspected it. He asked me to get someone I trusted to test the samples. The best I could think of was you and your parents." He pauses. "That's why I came here."

"Yes, you said that," she says, a bit harshly, and furrows her brow. "But you don't have the samples. Without them, all we have is third-hand hearsay about something that may or may not be HQV. That won't convince anyone. Not the way things are now."

"I have something else that might help. I'm not sure, but…" Ben pulls a slender pouch from underneath the sofa. It has two strips of cloth attached to it and is very dirty. "I didn't want you to wash it," he explains. He opens the pouch and pulls out a cotton handkerchief, folded over several times. He lays this on the floor and unfolds it slowly. Inside is another handkerchief. It has colorful hand-embroidery along the edges and looks heavily used. "This was Milly's," Ben says, his eyes glazing over. "Tilly made it for her." He does not continue. He just stares at the sad little handkerchief.

"Milly, who's Milly?" Eiko has to ask. "Another aunt?"

Ben does not respond. Eiko guesses that he needs time to sort something out and gets up to clear away the breakfast things. That done, she starts to make more coffee, then reconsiders and makes ginger tea instead. She brings two big mugs to the table and sits back down. Ben looks up. He is still not sure how to tell the story of Milly. He wants to make it a long story, stepping from one sweet, funny anecdote to another and ending on the magical story-telling that started after her first words. Until then, she had read so much, but said nothing. The words must have built up inside her. He imagines her standing at this very table, doing something with her hands, drawing perhaps, and looking sideways at Eiko. She would give Eiko one of her long-winded and fantastic explanations of why things work the way they do: why some flowers bloom in the spring, others in the fall, why bread needs to rise more than once, how forests can have no end. She would look very serious and be absolutely sure about her explanation. Ben had not been able to keep from laughing, sometimes. Her pique over his reaction would last only minutes. Then she would be off on some other tangent. Her words kept him sane after Tilly's death; they returned him to life. He wants to conjure Milly up, he realizes, and capture her, right here.

In the end, he does the exact opposite. He starts with the night they took her away. He describes the masked and hooded men grabbing the sleeping Milly from her cot in the back room. He describes Milly, awakened into a nightmare, kicking and screaming and shouting for him. He describes the poor woman, Milly's birthmother, running over from the neighboring farm, frantic and crazy, and clawing at Harold to let her by. All too clearly, he remembers himself, equally powerless, held by one of the oversized, anonymous helpers the priests employ for these jobs. With his eyes closed and his mind far away, he tells Eiko all of it. He had tried to keep it from the others, but Milly's nighttime attacks made it impossible. She would scream and cry and throw herself about, sometimes shaking violently. It was the headaches. Not every night, but more and more. He started sitting up with her, holding her upright while she slept to lessen the chance of an attack. But eventually, the attacks came no matter what he did. Her little face would distort with pain and her eyes would open wide with terror and incomprehension.

"I was never really in doubt," Ben tells Eiko as he opens his eyes. "It was just like in those old videos we saw. It had to be HQV." Eiko nods, but does not interrupt. "I suppose I knew they would come for her, and soon. I should have run off with her, maybe, but… There was nothing I could do. Once again… nothing." He lets the tears run freely, but continues talking. "Afterwards, I was a total mess. Again. I fought with Harold and Peter whenever I got the chance. I yelled at Daisy and Anne. The farm was in quarantine, but pretty soon they made me a special quarantine within the quarantine. That's when I had time to think. I knew I had to get back here. I asked for a priest, for Father Marius. My luck, priests are allowed to breech quarantine if they take precautions. He arranged everything after that. I'm pretty good at walking for a long time, it turns out. And at reading maps." Ben stops talking, looking completely spent.

"Oh, Ben. I'm so sorry." Eiko reaches across the table and rests her hand on his arm. She does not move closer, but stays like this, occasionally mumbling "you poor thing" or "that poor girl." Her facial expression shows that she means it, that she empathizes, but perhaps she feels that Ben needs his space.

After a while, Ben seems to recover. He goes to wash his face. When he comes back, he notices the small, partially unwrapped cotton package next to the sofa. He folds it up again, very slowly. Eiko gets up and goes to the kitchen area to fetch a sealable plastic bag. She returns and gently takes the folded cloth from Ben. He lets her.

"Will you take it to your parents?"

"Yes. I will bring it to them today. They will know what to do."

"So, if HQV is still out there," Ben starts slowly, "and there's still no vaccine or cure, then what?"

"Well, first we have to stop the hybrid madness—the 'let's all live together in peace and harmony' madness—*and* stop the war-hawks who will want revenge for the lies about HQV being eradicated."

"And then?"

"Fix the real problem. The original edits. Undo the block but retain HQV resistance."

"Is that possible?"

"I think so. We think so. And we're pretty sure we know how."

"Really?" He looks incredulous.

"Yes." She beams, proudly. "We're nearly there." She pauses. "Of course, not everyone thinks this is the right way to go. But once we have confirmation of active HQV, attitudes will change quickly. Our approach will be accepted and appreciated."

"You keep saying 'we'. Who's 'we'?"

"Rafi and me. We've been working on this for some time now. Well, it's no longer just us. We started it. Now there's a whole bunch of people involved."

"I thought Rafi wanted to work with plants."

"He did. He still does. But he also had a really clever idea about what might have happened with the HQV edits. We first talked about this-" she thinks for a moment "-a year and a half ago. After that, I looked extra closely at the edited sites. The regulatory loop involving the locus is actually quite neat..." Eiko tells Ben how they discovered the molecular logic behind the fertility block. She starts with the genetically modified peas, the empty pods and the Eureka moment. She describes her analysis of the HQV resistance loci, the complex RNAs and the distant chromatin effects. She tells him about transient rescue, genomic rescue and pre-em cells that finally behave as they should. Ben is an eager listener and asks questions along the way. Some aspects of their work she does not tell him about. It is too early for crucial details to be known by those who do not want them to succeed. The walls may have ears. The ears may not be friendly. As if an echo of this thought, they are suddenly interrupted by a loud and persistent ringing. Eiko points to the door.

"Are you expecting anyone?" he asks, softly.

"No. Absolutely not." She goes to the intercom. Ben stays out of view.

"Miss Ito Carr?"

"Yes. What is it? I'm working up here."

"Miss Carr. I'm Andy Fink, the building supervisor. An unidentified person was detected entering the building the night before last. I've been asked to check it out."

"Well, I've got no problems up here," she says. Why now, she thinks—why are they checking now?

"Are you sure? Your house computer is off."

"Yes, I'm sure." She pauses. "I always keep the house off."

"It's safer to have it on, Miss. Maybe I could come up and have a quick look around, just to be certain."

"Absolutely not. This is a studio apartment, not a castle. No one is hiding anywhere."

"Fine." He sounds displeased. "But let me know if you see anything suspicious. You never know with unidentified persons. They may be dangerous."

"I will," she says with a stiff smile and immediately cuts the connection. She goes back to the table and sips a bit of cold herbal tea while standing up. She looks at Ben. "Well, I guess you heard that. It won't be safe here for much longer."

"Should I leave?"

"No, stay. I'm going to go out for a short while. I'll bring this-" she picks up the handkerchief sealed in plastic "- to my parents. And I'll arrange for you to go to a safe place." She smiles, reassuringly. "I know exactly where to take you. I just need to organize the transport."

"Are you sure?"

"Absolutely. I promise to be back soon. Don't-"

"I know. I won't go out or answer the door or anything."

She smiles and moves a step closer to Ben. Then she turns around abruptly, picks up her coat and leaves the apartment.

After she has left, the place feels very quiet. Ben stays on the sofa, staring into space, his mind running through a multitude of memories. The problems of here and now have been passed on to other hands. The relief of that finally sinks in. He sleeps again.

*   *   *

"Ben." Eiko shakes his shoulder. "Ben, I think we should go."

He forces himself awake. "Sorry. I must have drifted off."

"Don't apologize." She smiles, briefly. "But we really should go. I have a pod ready in the parking garage."

"A hire-pod? Is that safe?"

"No, it wouldn't be. This is one of Celia's old ones. It has a pretty high grade shield, so we won't be tracked."

"Celia?" He sits up, quickly. "Is she here?" He looks worried.

"No. She is seriously AWOL. Didn't I tell you about that? I guess I didn't. Later—OK? She gave me access to the pod should I ever need it."

"And the... Milly's... Did you get hold of your parents?"

"I did. They send greetings—their love, in fact. My mother said she'd like to talk to you about Milly and... Later. They know where we're going."

"Good, because I realized that -" Ben notices the impatience on Eiko's face and stops talking. She has not taken off her coat and she is holding a second one over her arm. She hands it to him. The coat looks warm and slightly large for him—Paul's, possibly. He puts it on and follows her out the door.

A few minutes later, she starts the pod, transfers the coordinates and gets them smoothly into the afternoon traffic stream. Switching to automatic, she turns to him.

"This is a safe space. We can talk openly."

"Pretty cool space, that's for sure." He looks around in the ultra-modern pod. "How do you know it's safe?"

"Celia told me she had it wiped clean by an expert. After that, she registered it under someone else's name—or a fake name—I'm not sure. Plus the shield is almost as advanced as the one on her new pod."

"Wiped clean, fake names and advanced shields? I don't get it. And if she's missing, how do you..."

"I know how to contact her without raising any flags." He is about to interrupt but she keeps going. "She'll probably come visit at Rafi's. When she does, she can tell you all about her past exploits and her current shadow-existence."

"We're going to see Rafi?" A tentative smile relaxes Ben's face.

"We are." She smiles back. "He's got this great place: half botanical garden, half experimental vegetable plots and mostly still a mess. He does all his pet projects there. It's got several huts for living in—complete with whatever you need—and no surveillance, no drone access, total privacy. So once we're there, you'll be fine."

"Sounds awesome. How did he manage that?"

"I told you about him solving the timing-of-the-block mystery, right?"

"It sounds like it wasn't just him. You-"

"I've done my bit, too. But Rafi's insight was the key that did the unlocking. In exchange for this particular plot of land, he agreed to leave his intellectual contribution in the hands of some sensible people at the Reserve."

"What does that mean? He agrees not to publish? Or not to tell anyone? Or just that they get first jabs at exploiting it?"

"I'll let Rafi explain. We can't publish it, anyway. Not right now. The methods we've been using aren't exactly…"

"But how did he manage to get no surveillance?"

"He's a tough negotiator. He gets what he wants."

"Rafi? Really?"

"You'll find that he has changed quite a bit," she says. Ben looks worried. "In a good way," she adds quickly. "So, apparently, someone high up at state security—an old contact of his mother's or a CDL sympathizer—I don't know, exactly—agreed that Rafi could have a self-controlled X-something-or-other shield to cover the whole area." She looks thoughtful. "I think part of the reason he wanted it was to make it possible for Celia to visit. Not that this has happened much, as far as I know."

"Are they still…?"

"It's complicated." Eiko shakes her head.

"Still?" Ben widens his eyes in disbelief. They both laugh.

"So," Eiko starts once they have recovered, "now that we're sure we have no listeners, I'll tell you more about the fix and why we're so sure it will work. As I mentioned, we've been using some rather problematic methods to test the hypothesis and the fix." She frowns. "I don't mean scientifically problematic. They're just not entirely legal."

"Fire away. I'm dead curious."

"In the initial phase, we had significant help from the youngtwin community."

"It's a community now?"

"We've sort of been pushed together by the new regulations—and by people's reactions. People who previously had no problem with us suddenly see us like the New Eden priests do. As wrong—and possibly to blame—somehow." She sighs. He notices how natural the "we" and "us" seem, but does not comment. "We're not all navel-gazing discontents, like your old friend Leo." He lets that one go as well. "We just happen to be youngtwins. Anyway—Rafi's hypothesis predicted that youngtwins with an oldtwin from a previous generation should not be blocked, despite being fully edited. I got some of my friends to go off the patch that we're all supposed to wear and… get extra friendly."

"You were matchmaker? Madam of the youngtwins?" He exaggerates the joking tone, just to be sure.

"Neither. I just looked for the right opportunities." She smiles. "It's a small community, but I found several pairs who were happy to help. Once the first pregnancies were discovered, those directly involved were fined. But, naturally, no one will harm the babies. Babies are gold now. The authorities simply have to respond enough to make it clear that they are not condoning violations of the treaty with New Eden. They also shut down the youngtwin meetings. Thankfully, no one pointed fingers at me. No one pushed for it, either, as long as our misbehaving didn't escalate."

"This meant you could keep doing what you were doing in the lab."

"Exactly. So, we had our *in vivo* confirmation of the overall hypothesis. Half a year later, we had what we thought was the molecular explanation and a possible fix. I told you about that."

"You did. That was very clever."

"Thank you. So we ran the best test of it we had short of clinical work, namely humanized pig implantations."

"But wouldn't that give…"

"Yes, embryos that can't make it to term…" She shrugs. "We did only the minimum needed. It *could* be construed as disrespecting the spirit of the treaty, and this time at the Reserve, not by a group of naïve youngtwins. So it was done quickly and quietly."

"Did the people at the top know what you guys were doing?"

"Not at the *very* top, but quite a few at various levels below them. With the stakes being what they are, many find themselves willing to risk jobs, steep fines and house arrest to make things happen. Very few feel any real loyalty to the treaty."

"That's good."

"Yes. The tests were very encouraging. Pre-ems with the newly designed edits in place of the old ones and transiently rescued by injected RNA, newedits for short, developed completely normally. We broke out the champagne for that." She smiles and pauses. "Of course, we couldn't formally test whether they were resistant to the HQV virus, but everything we know says that they should be."

"So what's next?"

She does not answer immediately. She looks out the window instead. Traffic is less dense now the city center is well behind them. The light is softening. Trees seem to approach them at high speed. She looks back at Ben. Then she rubs her abdomen, slowly and deliberately. "What happens now is that we honor the original five, the original editees." She purses her lips and tilts her head. "No, that's not exactly it. We honor their parents, their mothers in particular, and their daring." She smiles uneasily, yet looks very determined. "It's the only way to show everyone that it works."

It takes Ben a few moments to process what Eiko is saying. "So you are…?" He starts to ask. She nods. He continues to shake his head. In disbelief, it seems. Then something else occurs to him, raising a sudden panic. He quickly moves as far as he can from Eiko, which is not very far in the snug little pod. "But I've been… Milly… You need to get away from me, Eiko."

"Don't worry, Ben," she says and reaches one hand out to him, the other still resting on her abdomen. "She is not a hybrid or anything like that. She is fully new-edited. She is not at risk."

"But you should… let me out." He tries to open the door. It does not let him.

"No," she says. "I'm taking you to Rafi's place and that's that. You brought us proof… We owe you."

"No, you don't. Don't be silly."

"*You* are the one being silly." She stares at him, hard, but he does not return the gaze. He keeps his eyes on the pod door. "I *do* understand, Ben," she says more mildly. "But you are being irrational." She waits. No answer. "Trust me," she finally says and turns to face straight ahead.

Ben continues to look miserable but lets go of the door.

They pass the rest of the trip in silence.

# 12

## Safe Haven

The dawn chorus woke Ben many hours ago. Now the sun is strong and he hears only a single bird, out there somewhere amongst the trees. He does not recognize the bird and does not try to. He is content knowing it is there and he is here, in the sunshine, on the porch of his little hut, shelling some of Rafi's fantastic peas for a salad. Their funny stripes still make him smile.

"I wonder what we're having for lunch today," Raphael says in a cheerful sing-song as he lets himself fall into the second chair. "Summer in a bowl." He answers his own question, eyeing the colorful concoction in front of Ben. "And, yes…" He leans back against the wall of the hut and sniffs the air. The smell of freshly baked bread is unmistakable. "A chunk of that glorious stuff, please." Ben smiles, as intended. They've made a habit of having lunch together every day. Usually they eat at Ben's hut. Although Raphael has been here for much longer, Ben's hut feels more lived-in and welcoming. Sometimes they are inside, a cozy space that looks like a country kitchen with a bed tucked in the corner. A pile of rescued books sits next to the bed; a small box of clothes lives underneath it. If the weather is nice, like today, they eat on the porch, which overlooks the rear section of the extensive grounds. They refer to it as "the gardens". Ben thinks the place should have a name. Raphael is not sure.

"Thanks for letting me stay here, Rafi," Ben says, somewhat serious. He has been wanting to say this for a long time. "It was just what I needed."

"Are you kidding?" Raphael moves out of the direct sun and looks at Ben. "I have the best deal by far: A hardworking and experienced helper, who," his face lights up in a mischievously grin, "since you are not *really* here, I don't even have to pay." Ben smiles. "*And* a fantastic cook," Raphael adds. He has

© Springer Nature Switzerland AG 2020
P. Rørth, *The Unedited*, Science and Fiction, https://doi.org/10.1007/978-3-030-34624-9_12

always eaten heartily, but now seems to appreciate the food more. He relishes the outdoor work, as well. It shows. He is no longer the pale, almost androgynous youth with a constant edginess. He is strong, tanned and seems completely comfortable in his own skin. Ben thoroughly enjoys his company. He has asked himself more than once why he so rarely thought of Raphael, let alone missed him, in the time he was away.

"Hmm." Ben smiles. "I see your point. Three months of eight hours a day-"

"Twelve hours, more like it." Raphael's expression turns more somber. "No, seriously, Ben. It's really great to have you back. You're a good friend and we missed you. We'll get the rest sorted somehow."

Raphael's words, especially the small 'we', reminds Ben of one awkward moment, right after he arrived. They were having a welcome dinner for him in Raphael's hut. The food was nothing special, but the wine was nice. Raphael had some bottles stashed away for a special occasion. His arm around Ben, he declared that this was a perfect occasion: a celebration of friendship. Eiko drank a little, Ben and Raphael the rest. At first, Raphael asked Ben one eager question after another about his time in New Eden. Ben's answers were slow and incomplete. He had let it all out when talking to Eiko and still felt drained from that. Ben's questions about Raphael's projects—the botanical garden, the experimental crops, the future plans—got long, enthusiastic and detailed answers, filling much of the evening. Eiko did not say much. Finally, Ben realized that he had forgotten to ask about the most obvious thing—and surely, it would be odd if he *didn't* ask. Although he tried for a light touch, it came out a bit bluntly.

"So, who's the father, Eiko?" he asked. "Or did you use an anonymous donor?"

"Of course not," she answered, irritably. "There's no such thing as anonymity when it comes to DNA."

"I am," Raphael said with a smile that seemed a bit forced. He looked back and forth between Eiko and Ben. Ben could not read him.

"It made perfect sense to do it this way," Eiko continued drily. "It is a joint project, after all." She looked in Ben's direction, unwaveringly. "And Rafi's got pretty decent genes."

"We're not a couple, if that's what you're worried about." Raphael added. "Romantically, I mean." This made Eiko swirl around and look at him angrily.

"Of course not," she said, snappishly. "This is… She is… much more important than that."

"I just wanted Ben to know," Raphael said, apologetically.

"Right," Eiko said, while continuing to stare at Raphael. Ben did not know what that was all about. He did not dare ask, either. Later, when Eiko had taken the pod back to the city and Raphael had shown him to his hut, he tried, unsuccessfully, to sort it all out in his head. Eiko had returned only a couple of times since then. She claimed that her absence from the city would draw unwanted attention to Raphael's safe haven. Ben was fairly certain there was more to it than that, but, again, did not dare ask. At first he thought that it was because of him. But, apparently, she had not spent much time at the gardens even before he arrived. He decided not to dwell on it but simply acknowledge that for him, the arrangement was perfect. He would be safe and Eiko would be safe from him.

In fact, he is more than safe. He loves it here. The gardens are green and peaceful. He works long hours, outside or in, doing things he enjoys. Some of the time, he is with Raphael. They talk, joke or are just quietly companionable. The rest of the time, he is with his friendly ghosts. Occasionally, he wonders whether he will ever feel whole again. But he no longer feels empty.

Back in the lunchtime sunniness, Raphael is still talking. "Don't you think?" he says, possibly rhetorically. He looks worried.

"Sorry," Ben says, "I zoned out for a moment there. What were you saying?"

"I was just talking about the elections. As expected, the PUP did really well. They will probably head the next government. No one seems to know what will happen once they do, or how fast. In the campaign they promised to open the domes and take definitive steps towards a future with New Eden."

"Just like that? Don't they know-"

"Right now, a lot of people simply want what they have on offer. They've fallen in love with the dome babies, especially the ones they know from the live feeds. That it's cruel for them to be confined to a dome for the rest of their lives is hard to argue against. More importantly, they see hope for children or grandchildren of their own. They are willing to take a lot of crap from the priests of New Eden to get a chance at that." He looks away. "I do sort of understand... And no war."

"But HQV—have they completely forgotten about that?" Ben asks, despairingly. "It's as horrible as ever." He told Raphael about Father Marius and his suspicions as soon as he arrived. Later, he also told him about Milly.

"Yuriko and Paul and their colleagues are doing what they can to warn the public. Unfortunately, they are considered fear-mongers now."

"Even with the evidence I brought back?"

"What was it Eiko said about the analysis?"

"They found a partial match to the original HQV sequence."

"Partial, yes. Of course, viruses evolve like everything else, so changes are to be expected. But it does make the argument less straightforward. People can continue to believe what they want. Since they never met your little Milly, I suppose the danger doesn't seem very real to them."

"Maybe I should step forward and tell my story."

"I don't know if that will help, but maybe…" Raphael nods a few times. "I've also heard that the Reserve might be closed down completely. Hopefully our 'new edits' will -" His wrist-link beeps. Transmissions inside the gardens are relayed via a special server with privacy filters, so he rarely gets messages. He fumbles with the tiny screen. "It's Eiko," he says, standing up quickly. "She sounds distraught. Something must have happened. She says she's coming out here—to stay."

"When?"

"She should be here in an hour. Assuming she doesn't have any trouble along the way." To Ben's surprise, Raphael freezes up. "What should we do?" he asks, sounding truly scared. "What if-"

"We should get her hut ready," Ben says, very calmly. "And after that, we'll go wait for her at the gate." He stands up. "We should bring a few things from here for her kitchen. And some flowers, I think, to make her feel at home. We also need to check-" Ben keeps talking as he goes inside and starts assembling glasses, plates and other useful items for them to take over. Lunch can wait. They need to keep busy for the next hour.

Almost two hours pass before the pod finally slides into the entry lock, with Eiko alone in the front. Raphael quickly flips the shield to let her in.

Eiko's roaming eyes and her pale, sweaty face betray the panic of the recent hours. When she gets out of the pod, she turns to Raphael immediately. He wraps her in a bear hug, holding her tight until she pushes him away. She gives Ben a tired half-smile and squeezes his arm. Ben turns to the pod to look for luggage, but does not find any. She notices.

"I didn't have time to go home first."

"But you are alright?" Raphael asks.

"Yes. I'm fine." She sighs. "It was just… madness, all of the sudden. I was leaving the lab when-"

"You were at the lab today? Don't you know…" Raphael stops himself. "Sorry," he continues. "I know I shouldn't tell you what to do. It's just… with the election results announced today, people might be extra…"

"It's OK, Rafi." She touches his arm. "And you're right. It wasn't very smart. I just needed to get away from my room and see other people. Early in the morning, everything was fine. But I should have picked a better time for leaving. There were people everywhere…" She lifts her hands, palms up. "I don't know who they were. Anyway, someone noticed my condition, despite my extra-loose clothing. Pretty soon, everyone seemed to know. They all turned toward me and stared at me. A few moved closer. Someone, a total stranger, touched me. She touched my belly. Then another. It was really scary. I screamed and shouted like a madwoman. I may even have hit someone." She sighs. "I suppose they were just curious. I've been thinking about it on the way out here. I was coming out of a building made famous by Reserve research achievements. Maybe people somehow guessed-"

"But you got away?"

"Yes. I ran back inside and took the tunnel to your old building. From there, I went out the back way, near the parking."

"If they open the domes—well, *when* they open-"

"Mayhem. I don't know how they expect to control it. People feel they know those babies. They'll want to see them, touch them and-"

"Do you think they'll do it soon?" Ben says, looking from one to the other. "Open the domes, I mean?"

"Possibly," Eiko says, nodding slowly. "It's the easiest campaign promise to keep. Openness and exchange with New Eden will take a lot more negotiation."

"The New Eden guys have no reason to hurry," Raphael adds. "They think time will only make everyone more desperate. But, for us, a delay is actually a good thing."

"Yes." Eiko takes a deep breath. "In eight weeks or so, we can show everyone that there is another solution—for real. This will change everything." She winces. "But right now I'm feeling a bit wobbly." In fact, she looks ashen. She staggers a few steps. Raphael and Ben each grab an arm. "I need to lie down for a moment." She continues. "Or maybe eat something."

"Why don't you take her to the hut?" Ben says to Raphael. "Rafi and I have made it nice for you," he adds, to Eiko, and slowly lets go of her arm. "I'll bring over some late lunch. And cake."

"Thanks, Ben." They say, in unison. Ben turns and walks rapidly toward his hut.

*   *   *

They settle into a modified routine. Raphael spends mornings finishing his paper on a new set of pea variants. The rest of the time, he is in the greenhouses or in the experimental fields. Ben is out there most of the day. There is plenty to do, this time of year. Ben is still the cook, as well, but now for four rather than two. Eiko seems to be getting bigger every day. She spends most of her time in her hut, finishing up her PhD thesis. She walks the gardens in the morning and in the evening, at a leisurely pace. Occasionally, she comes around to Ben's hut for a chat. They talk mostly of little things, and never for long.

Two weeks after Eiko has settled in, Yuriko and Paul come for a visit. They go straight to Eiko's hut. After an hour or so, Ben walks over to say hello and to ask if they will be around for dinner. They have to get back, they say, but thank him warmly for the offer. They seem genuinely happy to see him. Yuriko touches his arm. Paul pats his back. Ben is surprised at these gestures. Physical contact is not something they used to do much, as far as he remembers. Their interaction with Eiko has also changed. It seems gentler, more overtly affectionate, especially from Yuriko's side. They have brought along some of Eiko's favorite recordings. One of them, a collection of solo piano pieces that feel vaguely familiar to Ben, is playing in the background. Eiko seems to be cheerful and enjoying the parental attention. Ben is happy for her and appreciates the warmth in the room. But the feeling of being an outsider—an intruder, almost—is impossible to shake. He pushes the thought away and listens to the conversation.

"Is the Reserve still up and running?" Raphael asks.

"They haven't closed it yet," Paul says, "and we are trying to stay calm."

"The equipment is still there, and well maintained" Yuriko says. "So once baby Nee is born happy and healthy…" She smiles at Eiko.

"We have to use her arrival to the best effect," Eiko says. "We'll have to make it a public event of some sort." She looks intently at her mother and then at everyone else.

"Are you sure that's a good idea?" Yuriko looks concerned. "And that it's safe?"

"Yes and no," Eiko responds, carefully. "But we have to do it. There'll be a huge crowd for the opening of the first dome. We can attract an equally large crowd with baby Nee and the others. On a different day, preferably." She thinks for a moment. "The square in front of the Reserve would be perfect."

"I agree with Eiko," Raphael says. "We should announce the new edit strategy and results right there, in front of a supportive crowd. We'll explain the basics and we'll admit to everything we've done to test it, upfront." He looks at Eiko, who nods, then at Yuriko and Paul. "Including what wasn't legal."

"It *is* a good idea," Paul says, slowly. "Carpe diem."

"So to speak." Eiko smiles at her father.

"But…" Yuriko starts. She looks unhappy.

"People *need* to see that there's a solution that maintains our HQV resistance," Eiko says. "Knowing that, it'll be easier to accept the evidence that HQV still exists and that it's a threat to be reckoned with." She turns to Ben.

"That might be a good time for you to speak up, as well. Your experiences are unique and powerful."

"I think the scientists and clinicians need to hear from Ben first," Paul says, furrowing his brow, "and have a chance to ask questions. It's best if we come to a consensus before the public has a chance to rip it all apart."

"We are of course very grateful for the sample you brought us, Ben," Yuriko says as she turns to face him. "But you understand that there will be doubters. Especially now."

Sample, Ben thinks to himself, what a puny little word. He knows Yuriko means well, but does not trust himself to answer without bitterness, so he simply nods.

"Here," Yuriko continues, holding her wrist-link toward Ben, "let's exchange contact information. I'll arrange for you to visit the Reserve, meet some of the senior clinicians and testify about your... about what you saw."

Ben holds up his naked wrist with a shrug. Raphael hands him a slim e-stick and Ben passes it on to Yuriko. "Sure," Ben says, but a bit reluctantly. He gets the stick back and pockets it.

"If you *can* come in, that is," Paul says, carefully. "We wouldn't want you to put yourself in any danger."

"Officially, I'm still an unidentified." Ben raises his wrist again. "So it's a bit..."

"We need to do something about that." Yuriko says firmly. "Doesn't Marie know someone who can help?" She turns to Raphael. "Rafi? Have you asked her?"

"I should have," Raphael says. "But I never got around to it." He smiles at Ben. "I've been exploiting the cheap labor."

"It's been terrible," Ben says, returning the smile.

Yuriko's face shows confusion, then relief. "Speaking of Marie and government officials," she says, "I still think our best hope for public acceptance of the HQV threat is the medical committee—*if* they are given enough freedom to move around in New Eden and *if* we can get some of our people included."

"What's the medical committee?" Ben asks.

"Marie told me that parliament insisted on a high-level committee to evaluate health and safety concerns associated with an open union. The New Eden upper council seems to have agreed, but wants fifty percent of the seats, which is fine, and full control over any visits to New Eden, which is much less fine. We need someone on the committee who is willing to make contact with this Father Marcus-"

"Father Marius," Ben corrects.

"Marius, yes. Or with anyone else who can get them access to the so-called sanctuaries."

"Mama told me that last she heard, they still insist sanctuaries are holy places that cannot be visited by anyone impure," Raphael says.

"For once, let's hope she hasn't heard everything there is to hear." Yuriko sighs.

They talk of HQV and politics for a little while longer. When the talk turns more personal, Ben feels it is time to take his leave. He says goodbye and promises to stay in touch with Yuriko. As he walks back to his hut, he thinks of how much older Yuriko and Paul look. Four years—is that all it's been?

*   *   *

Ben wakes in the middle of the night. He sits up straight, sleep cast off in an instant, but cannot tell what woke him. The window shows the dark blue sky of a summer night. The clock shows almost three. The predawn chorus has not yet started, but he hears something else—a mechanical noise. He slips on some clothes and goes to the door. Stepping outside, the noise becomes slightly louder. Then it stops and is briefly replaced by faint voices. The noise resumes and he looks to where it is coming from, where the sky is a shade lighter. The gate is in this direction. As he gets closer, he can make out the lock and something dark moving very slowly in front of it. The voices start up again. He moves toward them, soundlessly, his bare feet suffering from the unseen stones and roots along the way.

"Of course you are."

He recognizes this as Eiko's voice. He sees a much taller figure, Raphael presumably, next to her. They are so focused on the person emerging from the pod that they do not notice his arrival onto the scene.

"Eiko—oh, my God," a woman's voice is saying "you are *huge*." She laughs. Ben recognizes the laugh. Celia. "Let me hug you again," she says and: "*You can wait.*"

"Right," Raphael says, presumably in response. He has his back to Ben.

Celia, still awkwardly hugging Eiko, is the first to notice Ben. "Oh, my…" she says with genuine surprise. "You're here?" She lets go of Eiko and walks over to Ben. Without hesitation, she gives him a hug as well. Then she moves him back to arms length and studies him. "Look at you! We thought you were dead or had become a monk or something."

"Not exactly," Ben says. "It's good to see you again, Celia."

"We're glad you were able to come, Cee," Raphael adds, a bit stiffly. She moves a few steps back.

So this was planned, Ben thinks. They didn't tell me. And they didn't tell Celia about me being back. He tries to ignore the sting of it. "I thought you were running from the law," he says to Celia, just to say something.

"I am, I am. It's quite tiresome," she says, but excitedly. He can see her outline and her hair tied back in a ponytail, but the expression on her face is obscured by the low light. "I just had to come see my best friends flaunting the law in a somewhat more radical way." She pats Eiko's distended belly. "And I get this wonderful surprise on top of it." She sounds truly thrilled as she turns to Ben again. He feels the rise of a flush and is happy for the cover of darkness. "Ben, Ben," she says, still incredulous, "imagine that—over the wall and back again. You must have quite some stories to tell. I want to hear it all."

The warmth stays in Ben's cheeks. He wonders whether this was a question and whether he needs to answer it.

"Can you stay for a while?" Eiko asks instead, addressing Celia.

"I can," Celia says. "But only if you're absolutely sure it's OK with you guys. I wouldn't want to-"

"Of course it's OK," Eiko says.

"The shield on this one is an X-27 beta. It never made it to commercial production but it's fully functional. So I'm practically invisible." She gestures at her pod, which, even at close range, looks somewhat like a large, dark smudge. "It was a parting gift from Victor." Ben notices Raphael flinch. "Sort of." Celia laughs briefly. "So, my dear friends, how about a drink for the weary traveler?"

"Let's go to my hut," Eiko says. "It's close by and I have a spare bed you can crash in. We'll get you settled properly later on." She turns to Raphael. "Rafi, could you bring an extra blanket from yours? And something to drink? I don't have…"

"Sure," Raphael says and walks off quickly.

"*Your* hut, Eiko?" Celia says. "I thought…"

"No, no," Eiko says. "It's not like that. Let's go inside, shall we? I'm getting a bit chilled." She starts walking toward her hut.

Celia follows her for a few steps, but then turns around to face Ben. "You aren't coming, Ben?"

"I'll catch up with you later," Ben says, now feeling the chill as well. "We can swap stories. I'm curious to hear what you've been up to."

"They haven't told you?"

"Only the bare minimum." He pauses. "I'll bring you girls coffee and some fresh bread in the morning."

"That sounds wonderful. Thank you."

"Special first day service." Ben smiles, even if she cannot see. "Don't expect it every morning."

"I won't." She walks up to him and plants a kiss on his cheek. "But I *am* happy to see you. I really am."

"Me too." He starts to move off. "Eight o'clock?"

"Make it nine, can we? I'm exhausted."

"Absolutely." Ben walks quickly and light-footed back to his hut.

He gets his feet warm again, eventually, but does not get any more sleep. He tries for an hour before giving up. By then, the predawn chorus has started for real. They love Rafi's place as much as I do, Ben thinks. He makes himself an early coffee and, wrapped in a blanket, sits outside to watch the dawn break.

He faces east but every now and again glances westward, toward pale pink reflections in the sky above the still-darkened woodland. The closest trees will shine a glorious green as soon as the sun hits them. The neighboring evergreens will wake up more slowly, wrapped in their dark green cloaks. Ben remembers the tour, on his very first day here. Raphael had been talking and talking, proud, excited and happy. Ben noticed some old trees in the corner of the forest section and went over to lay a hand on one of the giant trunks.

"Fantastic, aren't they?" Raphael said.

"They are." Ben patted each of the trunks. They reminded him of the trees at Margot's place. At the edge of the group he noticed a stump, recently cut and smoothened to make a table or a platform. He walked over to look at it more closely.

"Poor old thing. It fell down in a storm," Raphael said with heartfelt regret.

"It must have been…" Ben said and started counting as he moved his fingers over the silent growth rings, "over a hundred years old."

"A century and a half, actually. Imagine! A couple of the others may be even older. These ancients come from way back when, long before the crisis. What they must have seen…"

Ben went back to the nearest standing giant and gave it the best hug he could manage. "Unedited, I assume?" he said.

"Must be." Raphael smiled.

\*   \*   \*

"This is *so* cozy. I love it," Celia exclaims as they enter Ben's hut. She walks over to the long worktop. "It looks like you do some serious cooking here." She turns to face him. "By the way, the bread this morning was fantastic."

"Thank you." He smiles. "It was…" He stops and frowns. "It's not *serious* cooking, though—just cooking. We have all these amazing ingredients here. I think we've planted enough to feed a small army."

"So has Rafi given you the morning off to entertain an old friend?"

"It's not like that. I enjoy working in the gardens." He gestures toward the door and the sunshine beyond. "My small contribution to the green bounty."

"He ran off this morning saying there was so much that needed attention right now. I don't want to keep you from it."

"It can wait. Eiko needs to finish her chapter. Rafi needs to… whatever. I need to hear your story." He smiles. "Hiding from the law, in contact with Leo *and* with his Dad? I am totally mystified."

"They really didn't tell you?" She looks skeptical.

"Eiko said that you'd tell me about it yourself. She seemed to know you'd show up at some point."

"Hmm." Celia frowns, then sighs. "Well, it's a long story."

"Even better," Ben says. "I'll make some more coffee." He gestures to the door again. "Let's sit outside."

Celia looks at the old-fashioned coffeepot and the grinder that he starts filling. She shakes her head with an indulgent smile and goes outside.

When Ben comes out a few minutes later, Celia is sitting in full sunlight with her eyes closed. She looks content, serene, and as beautiful as ever. He moves toward the table slowly, careful not to disturb her.

"I'm not asleep. I'm just enjoying the morning rays."

"It'll get too hot later."

"It already is." She turns her head and opens her eyes to look straight at him. He is surprised at the jolt he feels. She smiles. Knowingly, he imagines. She moves her chair into the shade. "Story time," she says, "as promised."

In the following couple of hours, as he and Celia talk, Ben gradually comes to appreciate how much she has changed. The toughness and confidence is no longer being tried on and shown off. It's just there. It fits. He wonders how much he has changed and how.

Celia starts by telling him how she met Victor Huang. She describes their lunch and Leo showing up.

"So Leo actually works with his father?" Ben sounds incredulous. "In the company?"

"He did—now they… I'll get to that later." She makes a dismissive gesture, then frowns. "Naturally, I was surprised to see that. I remembered how aggressive he had been about his father that summer. He still hates him, it turns out, but now keeps that well under wraps. The fertility block and the political crisis that followed have changed a lot of things."

"I guess the danger of half-siblings, of 'normal' siblings, is gone," Ben muses. "Leo couldn't stand the thought of that."

"That's part of it, I suppose. But things also changed for Huang Shields. The company was successful already, but the threat of war with New Eden as well as the general unrest made sales of shields shoot way up: the big public ones got upgraded and smaller, private ones really took off."

"Like on your pod."

"Yes, and the one Rafi has here." She gestures to the surroundings. "Anyway, with the expansion of the business, Victor could no longer oversee everything himself and Leo saw an opportunity."

"But how did *you* get involved with Huang Shields? And doing what? You said the meeting was about security."

"It was." She pauses. "You know I started a small company?"

"Eiko mentioned it. Something about computer models... simulations? But I didn't get what they were for. She said it was just you and some super-shy, nerdy partner."

"Not a million miles off." She smiles. "The models were for all kinds of things. The approach was based on treating people and AIs as discrete units that influence one another's behavior, individually and collectively, with realistic response parameters and response times. It was called Response Modes. There was never any partner, though. That was just a convenient fiction. I did the business side *and* the actual modeling and computing."

"Wow."

"There are some advantages to being a hyper."

"Being what?"

"A hyper." Celia sees that Ben does not understand. "Hyper-selected and hyper-edited... Let's just say the allele-combinations worked out well in my case."

"OK." Ben looks a bit uneasy. "I've just never heard the phrase before."

"It's new." She shrugs. "I don't mind. I mean, the word, 'hyper', labeling us... That's stupid, of course. As if there's some sort of magic line that makes us different from everyone else." She scoffs. "I guess people always need to... whatever. The point is, I'm perfectly happy with who I am. I enjoy being competent—extremely competent in some areas. I've had lots of opportunities and I've made all my own choices. What more could a girl want? Seriously." She does look serious—as if it matters to her that he understands. He nods.

"So," Ben says after a short silence, "back to the story. What did you actually do for Huang Shields?"

"I helped with strategy." She pauses for a while, her expression shifting from thoughtful to worried to determined. "They wanted to optimize sales of private shields, where their profits were the highest. Victor asked me to explore what level of uncertainty and fear, and which kind, would make as many individuals as possible buy the best shield they could afford rather than being complacent and expecting the government to take care of everything or, alternatively, becoming paralyzed with dread and hopelessness."

"So they were trying to take maximum advantage of the situation. Not nice, but I suppose that's business."

"That's how it started. Had it only been that, I could have dealt with it. But the next level was worse. Starting with specific templates of theoretical rumors, he asked me to determine which rumors, and coming from where, would generate the "desired" level of uncertainty and fear."

"And these rumors…"

"Started appearing exactly where and when my predictions said they would have the maximum impact on sales."

"So you helped Huang Shields drive people into a fear-fueled buying frenzy?" Ben cannot keep the disapproval from his voice.

"Not for long. I stayed close to them for as long as I needed. As soon as I had enough evidence, I exposed their double-dealing. I helped drive them out of business and make them subject to criminal charges."

"Double-dealing? Who were they-"

"New Eden, of course. They were selling to the enemy! They sold specs of everything that was being developed, complete with timelines and expected distribution. Private shields *and* public shields. They even sold critical parts for the shields themselves. Somehow, these were smuggled across the border. And somehow, the priests found the resources to pay for it all." She is now speaking fast and with emphasis. "Do you remember Father Elias?"

"Vaguely. He was at that transit center." Ben thinks. "Yes, of course… He was the one who told me about my status *and* about the-"

"Leo made some sort of a deal with him. They were in regular contact, those two. Actually, I think the whole thing started when we were at the transit center with you."

"I didn't notice."

"It was probably after you left." She tilts her head, thinking. "Yes, that would fit. Father Elias told us about you having crossed over and, after that, he told us about the fertility block. Leo must have worked out the most likely consequences, personal and business-related. He's not stupid, the boy." She shakes her head. "Just moderately deranged."

"So it was Leo who handled this double-dealing?"

"Yes. And he made an absolute fortune doing it. I'm not sure whether Victor ever knew what was going on. At one point, he must have suspected something. But Leo was clever; he hid it well." She shrugs. "Anyway—everyone assumes Victor knew."

"Because he was in charge of the company?"

"There *is* that, of course... but also because Leo is his youngtwin."

"Meaning he would know what Leo was thinking—and doing? People don't really believe that, do they?"

"People believe all kinds of shit." She shrugs again. "With Leo it was an open and shut case. Given the specifics of the situation, it was considered treason. There was no direct proof against Victor—proof of amoral conduct in deliberately spreading disruptive rumors, yes, but no proof of actual treason. They convicted him anyway."

"So they've been tried and sentenced?"

"Yes. They got lifelong high-security detention—that's essentially house arrest in one of the ultra-tight mini-domes. Not one of their own domes, obviously." She smiles, slightly mischievously. "Also, they are not allowed interactive access to any part of the net. They only have incoming signal— entertainment and so on."

"I was just about to say that was a mild sentence, but-"

"It gets better," she says and pauses for effect. "They serve their sentence together. It's considered more humane that way." She laughs, sharply. "I don't know who was more freaked out by that, Victor or Leo."

"Oh, shit." Ben smiles. "That must be a teeny bit tense."

"Explosive."

They both shake their heads, contemplating life in the Huang mini-dome.

"And you?" Ben finally asks.

"For my involvement in the spreading of misinformation, I got a large fine, which I paid immediately, as well as five years high-security detention." She looks over at Ben and adds, with a touch of drama: "*Not* with the Huangs." He nods. "Both Victor and Leo swore revenge on me, loud and clear. I had gotten to know them pretty well, so they felt betrayed. Rightfully, I suppose." She sighs. "My sentence seemed unreasonably harsh, especially since I was the one who exposed the whole damn thing. The judges must have thought I was involved in the treason part, somehow—that I had made money on it and stashed it somewhere.

"And *did* you make money on it?"

"No!" She looks shocked. "Of course not. How could you even think that?"

"I don't know…" He looks down. "Your fancy pod. And Eiko tells me you have this secret place with-"

"Yes, I've made a lot of money. But all legitimately. Not by selling secrets." She keeps staring at him, her anger real. "Anyway, I'd never want to have *anything* to do with those idiotic extremists."

He is about to ask who the idiotic extremists are when he realizes he knows the answer already. He decides not challenge her description. "I'm sorry, Celia," he says instead. "Really. I'm sorry I said that. I didn't…"

"Apology accepted," she says. But her voice is stiff and she looks at Ben only briefly. She is obviously still upset. She turns and gazes toward the far-off trees. Ben stays quiet. Gradually, her face relaxes. "Because of the threats from the resourceful Huangs," she continues, her voice back to normal, "my mini-dome had to be extra secure."

"Yet here you are." He tries to sound jolly.

"As I said, I'm quite good at certain things." She smiles, but sadly. "Some people I trust will alert me if the Huangs ever get out, or if they get outside contact privileges. That's my real worry. Keeping out of sight is prudent, anyway. The long arm of the law isn't trying too hard to grab me, as far as I can tell, but it's best not to tempt fate."

"So will you go back to your secret place—or will you keep moving around?"

"I'll probably stay here for a little while. I'm looking forward to being First Aunty. If Rafi will let me."

"Of course he will. Rafi loves you. He always has."

She smiles, enigmatically. "Maybe."

"But Eiko said…" Ben starts. "Now I'm officially confused." He scratches his head. "Are you and Rafi still—whatever it's called—seeing each other?"

"Lovers is the correct term." She looks straight at Ben, who feels the blush rising and looks away. "No," she continues, "not anymore. It's better that way, for both of us. He's clearly much happier now. He's more relaxed and confident." Her smile is not quite convincing.

"But that's because of this, isn't it?" Ben gestures to the gardens. "And because he cracked the all-important timing puzzle."

"He did, didn't he?" Her voice trails off. "You are probably right, Ben" she continues, more briskly. "Rafi's blossoming has nothing to do with me. But, in any case, we are not good for one another."

"My point is this," Ben persists, "Eiko seems to think that you two are still—together." And Rafi *is* behaving oddly today, he thinks. It's not just my imagination.

"That's ridiculous," Celia says, gruffly. "I'll have to set her straight, then." She is looking at the trees again.

"Did you know…" Ben's mind is racing, trying to remember who said what, exactly. "Do you know that Rafi…" He falters.

"Is the father of baby Nee? Yes, of course. That's why I was so surprised that Eiko is sleeping on her own." She shakes her head. "They even asked my permission, beforehand." She lets out a brief laugh, like a snort. "Now *that* was strange—both them asking and my having to answer it. I've always told Rafi that he is his own man."

"They both say that they are just friends. And someone had to be the genetic father. Rafi was a good choice."

"He *is* a good choice. But it can't be that simple."

"Maybe not."

"Oh, Ben," she says, tilting her head. "I didn't mean to discourage you. I know you and Eiko…"

He stares at her, shocked. Does she really not remember? Four years ago? That stupid, starry night? Her eyes are full of sympathy, bordering on pity. He hates it. He looks away.

"What kind of name is that, anyway? Nee?" He asks after a while.

"Short for New Edit, I think."

"Stupid name."

"I agree." She shrugs. "Hopefully they'll come up with a better one when they see her."

Ben makes a sound somewhere between a huff and a grunt.

"Which reminds me," Celia says after another stretch of silence, "I'd better go check on Eiko."

"And I'd better go help Rafi." He pushes back his chair and picks up the coffee mugs.

"Will you tell me about your adventures sometime?" Celia says, light in tone.

"My adventures?"

"Your time in New Eden. I have to admit that I'm curious about that crazy place."

"I only saw one small bit of it."

"Still, I'd love to hear…"

"Some other time, OK? I really should…"

"Of course. That's fine," Celia says as she gets up from the chair. "Thanks for coffee and—for everything."

"You're welcome," he says, irritated by the generic comment—what does that even mean: "for everything"?

He watches her walk toward Eiko's hut, barefoot and summer-dressed. She moves with ease, confidence and an unmistakably female smoothness. Desire tugs at him, but dulled, like an echo. He is doubly disappointed with himself.

* * *

It is mid-afternoon. Raphael and Ben are working together in the bean section. This is a large area, with many new varieties being tested. They are near the southwest corner.

The headache hits Ben like a hammer. He grabs his head and winces.

"What's wrong?" Raphael asks. He is close enough to see Ben's distress.

"Headache. I didn't drink enough water, I think."

"Out of the sun you go!" Raphael orders. "We'll finish this tomorrow."

"Sure, Boss," Ben says, with a forced grin. He grabs a large cup from a nearby table and moves toward the closest water tank. Dunking his head in is like a shock to the system. It helps, for a short while. He stands up straight and blinks a few times. He pours filtered water into the cup and drinks it, slowly.

Moments later, Raphael sees Ben walking, somewhat unsteadily, toward his favorite group of ancient trees. He can see Ben easing himself into the shade and support of one of the old giants. He cannot see Ben's terrified face.

# 13

## Summer Rain

"Is that her?"

"It can't be. They'd zoom in for that."

Ben and Celia stare at the silent screen. The camera is slowly panning the crowd. It is evening, but still light. The square is packed. Everyone is waiting.

"So many people," Ben says, "even in the rain."

"It's a big deal—history in the making," Celia says. She frowns. "How can you not have sound on this thing?"

"It's one of Rafi's cast-offs."

"I'm sure he can do better than that!"

"It was fine when I got it. I spilled some-"

"Shush! She's coming out."

The camera zooms in on the front door of the Reserve, which is opening slowly. Once it is fully open, Eiko steps outside with a bundle in her arms. She is smiling broadly. The screen splits in two. One of the cameras pans back out, the other zooms in on baby Nee's face. The angle is not perfect and neither is the focus, but it is enough. As soon as the image goes up on the big screens, the crowd goes wild. Arms rush up in excitement. People hug their neighbors. The first camera zooms in on faces in the crowd. Delight, relief, disbelief—everyone is emotional.

"How did they know to be there today?" Ben asks, eyes glued to the screen.

"Net rumors had it happening soon, so people started camping out on the square, to keep a look-out and to secure a spot. There were people keeping watch at four this morning, when I took them in. They must have seen her."

"How was she?"

© Springer Nature Switzerland AG 2020
P. Rørth, *The Unedited*, Science and Fiction, https://doi.org/10.1007/978-3-030-34624-9_13

"Alternately screaming and panting." Celia widens her eyes. "Fine, I suppose. I was focused on getting us there as quickly as possible."

"And Rafi?"

"He was panicking. I'm glad Yuriko and Paul were at the other end. Yuriko took charge immediately."

Ben points to the screen. "Look, it's baby Nee again. Eiko is holding her so we can see her face better."

"Newborn babies are kind of ugly, aren't they?" Celia says with a big smile.

"I suppose." Ben smiles as well. Then he points again. "Look, she's got hair already. A whole head of dark hair."

"Ah… And look at those chubby little hands, would you? Waving at us."

"Eiko is stepping up to the microphone. And there's Rafi."

"Two steps behind. He looks more normal now."

"Eiko is talking." Ben states the obvious once again, but Celia doesn't mind. They watch Eiko's lips move. "She must be explaining everything, don't you think?"

The camera stays on Eiko while she speaks. Her expression is serious, but she ends with a radiant smile. She gestures at Rafi. He steps forward and speaks into the microphone for a minute or two. As he steps back again, one of the cameras pans down to the area in front of the stage. There are two solid rows of police at the very front.

"Are they going to arrest them, do you think?" Ben asks.

"Not in front of all those people. I imagine the crowd is hard enough to control as it is."

On screen, Eiko turns around and the door to the Reserve opens again. Two women come out, both bearing bundles and both smiling. They look considerably older than Eiko.

"Eiko tried to tell me about the other mothers on the way in—between contractions." Celia points at the screen. "I think this one is Amita—Amita Singh, Eiko's PhD advisor. The other one is also a professor. Caroline something. I didn't catch her last name."

"So all three of them have new edit babies? At exactly the same time?"

"I'm not sure when. You'll have to ask them. But they wanted to announce it together. And since Eiko and Rafi-"

"-made it all possible, they had to wait for them."

"Yes. They told me it was important that baby Nee wasn't the only one. She might have been seen as an exception, a fluke. There was also something about Eiko being a youngtwin. I don't know why that matters."

"She is, of course, technically. But it shouldn't matter in this context, given that-" Ben stops when he sees that Celia's attention is drawn to the screen again. "Anyway, it's smart. This way there's no doubt," he finishes.

"Show us baby Nee again." Celia leans into the screen.

"You're being quite the First Aunty."

Celia smiles. "Absolutely."

"Are they coming back here?"

"Eventually. Assuming they don't get arrested." She frowns. "The plan is for Yuriko and Paul to take them home first and then bring them here when it's safe." She looks at Ben. "With so much attention, I don't think either of us should go in to see them. Even with the X-27."

"I agree."

"Maybe I shouldn't even be here when they return. Who knows how much press there will be? I don't want my presence to confuse matters."

"You *have* to be here. They would be upset if you disappeared before meeting baby Nee—Eiko especially. Just stay away from the gate."

"You are right, Ben." Celia looks at him with fond appreciation. "You are a good friend. I'm so glad you came back." After a while, she shifts her gaze to the middle distance and adds, with some sadness, "but it will all change now."

"It will," Ben says, heavily.

* * *

Three days later, they are having breakfast in Eiko's hut. It is still raining out. Inside it is damp and crowded, but cheerful. It is the first time the five of them are all together. Ben has brought over various baked treats, more than they can possibly eat. He is hovering between the coffee machine and the table. Eiko and Raphael both look very tired, yet somehow alert. Celia is the most energetic. "So, explain. The two other babies…" she quizzes Eiko.

"Amita had her little girl three weeks ago. She is the cutest thing… Caroline was just a few days before me. The final two still have some weeks to go. Now seemed the best time to announce: Three perfect little girls, each with ten fingers and ten toes."

"Three girls?"

"That seemed appropriate." Eiko smiles. "The next two are boys."

"And all are new edits?"

"Well, Caroline's baby is…" Eiko looks from Celia to Ben and back again. "You didn't see the live feed? I explained all this."

"Yes, but no sound," Ben says. Eiko looks puzzled. "Long story," he adds.

"For another day," Celia says.

"Clumsiness." Ben shrugs, but smiling.

Celia addresses Eiko again. "You were saying—Caroline's baby? Is she a hybrid?"

"We should stop using that term," Raphael says from behind baby Nee. He is on burping duty while Eiko has breakfast. "It's not very nice. Call it mixed parentage."

"Sorry. You're right," Celia says to him. She turns to Eiko again. "So, what's the deal with Caroline's baby?"

"Both she and the baby's father are edited, the same as everyone else. That was the whole point," Eiko says. Ben wonders if she will look at him. She does not. "But their daughter was rescued transiently, by the RNA only."

"Why would they do that?" Celia asks. "The block will just reoccur in the next generation, won't it?"

"It probably will. But she, or they, wanted it this way—without introducing a permanent genomic change. They felt it was less of a risk. Anyway, there's no reason to think the rescue can't be repeated in the next generation."

"It's actually good for the program," Raphael says. Just then, baby Nee burps. A white mess lands on Raphael's shoulder. He makes a face. "You explain, Eiko," he says and wanders off toward the rear of the hut.

"It means there's more than one approach for anyone wanting to get pregnant now. If someone is not comfortable implementing a set of freshly minted edits, they can choose transient rescue instead."

"But HQV resistance is maintained in both cases?" Ben asks.

"Absolutely," she answers firmly, turning her attention to him. "Either by the new set of edits or the old set. The coding output is exactly the same."

"So—it's a choice? We won't have a repeat of the situation back at the crisis, where everyone had to follow the same protocol?" Celia asks.

"Yes, it's a choice. I guess we've learned a thing or two from our brush of Hubris." Eiko pauses, letting the comment settle. "I've been told that the sign-up is about equal for the two approaches. And it's already overwhelming. The Reserve will be very busy, for a long time."

"If they are allowed," says Celia. "Isn't it still illegal to do this?"

"Yes, in principle, but it won't be for long. Marie and Anton came by the house yesterday. Marie told me the PUP people are busy changing their tune. They want to stay in power."

"With a platform that seeks unification with New Eden?"

"I don't know what their platform will be. That's politics. What *is* clear is that New Eden will no longer be able to dictate the terms. There's only one thing people really care about now."

"More babies like baby Nee!" Celia exclaims and holds out her arms.

Raphael has just returned in a different T-shirt and with a sleepy-looking baby Nee in his arms. "If they are smart, part of their platform will be flexibility," he says. "On the biomedical side, that could include a serious effort at combatting HQV directly." Celia and Eiko look at him with surprise. Ben's expression is darker, but no one is looking at him. Raphael shrugs. "It's sensible, isn't it? It could also serve as an olive branch to New Eden." He hands the baby over to Celia and sits down across from Eiko. Pouring himself a cup of coffee, he contemplates the table full of food.

Celia makes baby-cooing noises for a while. "Woopsie!" she suddenly exclaims, accompanying a jerky movement. Eiko immediately looks over, alarm written all over her face. But, apparently, Celia has a good grip on the baby. Eiko bites her lip and turns back to her plate.

"Do you want to hold her for a bit?" Celia looks toward Ben and starts to get up from her chair.

"No, no," Ben says. "You hold on to her. I'm…" His expression goes from alarm, to pleading, to sad. He looks out the door, into the rain.

Celia sits back down, eyes on the baby.

Raphael and Eiko exchange glances.

"Ben," Raphael finally says, "we have a favor to ask you." He looks at Eiko and gestures for her to continue.

"We'd like to call her Milly," she says. "Baby Nee was just…" She searches Ben's face for a response. "But only if that's alright with you, obviously. Maybe it's too…"

Ben swallows. "That's…" He swallows again. "Of course it's alright."

"If you'd rather we didn't…"

"No, no, it's fine. It's more than fine, actually." He tries to smile. "Milly is a good name."

"It is," Eiko says.

The whole table is quiet for a few moments.

Eiko gets up and walks over to Ben. "Milly's suffering was not for nothing," she says, with an awkward formality. "You should know that, Ben." She takes one of his hands. He wants to pull it back but resists the urge. "What you did has made a huge difference," she continues. "The evidence that you brought back and your testimony—my mother told me a bit about it—it convinced scientists *and* politicians that HQV is still a real threat." She looks him straight in the eyes. "You did something important. You and Milly. You saved a lot of people from a terrible disease."

He smiles and nods in response but finds no words that fit. Eiko finally lets go of his hand.

\* \* \*

"Yuriko? It's me, Ben."

"Ben. How is everyone out there? How is baby Nee? Is she alright?" Yuriko is speaking fast. "And where are you? There's all this stuff in the background, I can't…"

"They're all fine. Having a nap, I think. I'm in one of the greenhouses."

"Are you coming in for the festival tomorrow?"

"No. I can't, I-"

"But Ben," she pleads, "you should be there. You're such an important part of-"

"I can't," he repeats, a bit abruptly. He takes a deep breath and continues softly: "I'm afraid it's time."

"No…" Yuriko's face on the screen fills with distress and sympathy. "Are you sure, Ben? Are you *absolutely* sure?" She rushes on. "They're talking about starting up a new approach to combat the infection. It might-"

"It's too late for me."

"But…"

"I *am* sure," he says, nodding slowly. "The headaches are getting much worse. And they come at any time."

"Do you have those-"

"They don't do enough—or they knock me out." He looks down. "I don't want them to see me like that."

"I understand."

"Do you need anything more from me?"

"More?"

"Blood samples. Scans. Measurements."

"Ben… you've been absolutely fantastic." She looks both sincere and serious. "We owe you so much. We would never have known—not for sure, anyway—if you hadn't come forward."

He nods a few more times.

"But do you need anything?"

"Another round of scans? And blood samples? If you're sure you don't mind."

"I don't. I prefer to be of use."

"Oh, Ben. That's…" Now she looks very upset. "Thank you," she finally says.

"So will you contact Sunset and explain everything? I'll get a hire-pod to take me there tomorrow, when everyone else is downtown."

"Are you sure about Sunset? It's all old people, you know."

"My mother used to work there."

"She did, didn't she?" Yuriko looks almost cheerful for a moment, but loses it quickly and starts blinking rapidly. "She was a strong woman, your mother."

"Yes, I believe she was," he says, slowly and thoughtfully. "Some of the staff might remember me," he continues, more rapidly. "She took me along a few times."

"She did?"

Ben nods, but does not expand on it.

"If you are sure, Ben. Absolutely sure."

"I am."

"Then I'll contact them. But tomorrow we'll be-"

"Tomorrow you and Paul should be with Eiko and Rafi and baby Nee, celebrating. I'll take a hire-pod."

"I'll take him," says a voice from behind Ben. He turns around quickly. Celia is standing in the open door, an umbrella by her side.

"Who's that?" Yuriko says on the screen. "Who's there?"

"It's Celia," Ben says, as Celia closes the door behind her and moves toward the screen. "You remember Celia from our class, don't you?"

"I'll bring him to Sunset," Celia says, facing Yuriko via the camera. She does not add any explanation.

Yuriko seems too shocked to come up with any questions for Celia. She rings off a moment later. Ben and Celia do not move. They hear the rain falling on the glass roof.

"How much of that did you hear?" Ben finally asks.

"Enough. But it doesn't matter. I knew anyway." She looks straight into Ben's eyes. "I think you are very brave."

"Not very brave, just very sick." He takes a deep breath. "How long have you known?"

"I guessed some time ago," Celia says. She reaches out a hand and takes hold of his sleeve. "When they came back with baby Nee, I was sure." She smiles, softly. "Baby Milly."

"It's easier if I take a hire-pod."

"Don't be an idiot! I'm taking you." She sounds almost angry. He does not respond. "You don't go to Sunset on your own," she continues. "Not unless you have no one who loves you."

"But I..." Ben falters. He looks at Celia, but avoids catching her gaze again.

"Don't make this more complicated than it is. I'm taking you and that's that. If you prefer, we won't tell Eiko and Rafi. We'll send them texts explaining everything after the festival."

"I prefer that," he says. She nods. "Do you think they know?" he adds suddenly. "About me?"

"I doubt it. They've got too many other things on their minds." She pauses. "Are you sure you don't want a proper goodbye?"

"I am." He straightens up. "But won't they miss you at the festival? It's a big day for them."

"I wasn't going to go anyway."

"Of course not. Stupid me. A bit too exposed." He tries a smile.

"It's not my scene, anyway." Celia looks around and gestures at the plethora of plants with a bemused expression. "None of this is. It's time for me to move on."

"What will you do?"

"I'll be fine. I've created a solid new identity and I'm ready to jump into it full time. I just can't go backwards."

"So this is it?"

"Yup." She purses her lips.

"Celia?"

"Yes?"

"Did it bother you at all? Eiko and Rafi? Baby Nee?"

She smiles, wistfully. "Of course it did." She shakes her head very slowly. "I may be hyper, but I'm only human."

"And?"

"That's life."

"So you are not going to be First Aunty and all that?"

"As I said: It's not my scene. They don't need me and I'm not sentimental. Nor am I a masochist."

"And yet you want to take me to Sunset?"

"That's completely different."

*　*　*

"Dr. Eisner," the nurse yells down the corridor. "Dr. Eisner. He's here." She looks at Ben and adds more softly. "Bella's little boy is here." She is a large woman, with a moon-shaped face and a generous smile. She takes both his hands in hers. She smiles kindly but briefly at Celia, who takes a step backwards. "We will take good care of you, Ben," she says to him. He looks into the deep pools of her eyes and knows that he trusts her completely.

# Scientific Appendix

## Manipulating the Human Genome: Technology and Choice

As I was mid-way into writing the accompanying novel, *The Unedited*, an unexpected story hit the news. Dr. He Jiankui announced that he had enabled the birth of a pair of genetically edited twin girls by using a powerful new biotechnological tool called CRISPR-Cas9 to manipulate the girls' genomes at the early embryo stage. Although the report was not published in a peer-reviewed scientific journal, the claim seemed likely to be true. The experiment was quickly judged illegal and the scientific community largely condemned it (Cyranoski and Ledford, 2018). Unsurprisingly, the general press reacted as well. Whether plants or animals, genetically modified organisms, or GMOs, remain contentious for many. But producing genetically modified human beings? That is about as controversial as biotechnology can get. I will return to the specific experiment later, but the key message is this: Human genome editing is no longer just a matter of speculative fiction. It is here. And the issues this raises call for serious consideration from the scientific community and from society as a whole.

Does the advent of the means to edit the genome mean that "designer-babies" are right around the corner? I doubt it. Our biological know-how is not advanced enough for wide-ranging human genome editing, let alone "design", to make sense—even to the most eager. This may change, of course, but only slowly. In addition, public perception of editing is unlikely to be

© Springer Nature Switzerland AG 2020

P. Rørth, *The Unedited*, Science and Fiction, https://doi.org/10.1007/978-3-030-34624-9

favorable and societal regulations will most likely continue to outlaw it. This may change, as well, and potentially more quickly.

Public perception is likely to be guided by two important, but fundamentally different, concerns. The first, as with any new technology, is safety. Do we fully understand the consequences of editing the genome? Recombinant DNA technology and GMOs have been discussed in scientific and public forums over the past 50 years, resulting in policies and regulations controlling their use. Turn to human genomes and the stakes are much higher: genetic fixes may alleviate life-long suffering or early death, but mistakes could also be devastating. Is the risk worth the reward?

The second concern is reproductive choice. The rapidly advancing power of genome-based diagnostics is making choice a significant ethical issue facing us even today. This is a sensitive subject. We consider each individual human being to be unique and inviolate. From this perspective, it may feel wrong to choose or alter what someone *is supposed to be*. The technology to select embryos based on information about their genomes already exists and some prospective parents use it to ensure that their children do not carry specific disease mutations. In western society, views on embryo selection are likely to range from abhorrence of the very idea to an acceptance that gene-based selection has a place, given the right circumstances. Eliminating a known disease-causing mutation is a long way from designer-babies, but the newly available potential for at-will genome manipulation does bring a whole new dimension to these ethical issues.

In the novel *The Unedited*, I have assumed that the technological hurdles to human genome manipulation have been solved. There are number of factors implicit in this, but, as it stands, the underlying technologies are not far from the mark. Reproductive technologies, which generally depend on in vitro fertilization (IVF), are well established; for non-human mammals, this includes cloning and twinning. Genome sequencing is now fast and inexpensive, and the tools for individualized genome analysis are advancing rapidly. Precise genome editing is now also possible; there are some caveats, but technical improvements will continue. Less certain is to what extent we can predict the biological outcome of specific genome manipulations. The novel explores these areas in a fictional format. In the following, I will discuss the factual context: what is currently possible and what might be possible in the not-so-distant future.

# Reproductive Technologies

## IVF

The first baby conceived by IVF, Louise Brown, was born in 1978. Since then, the technique has gained wide acceptance and million of babies have been conceived this way. The original purpose of IVF was to enable people who, for one reason or another, have a difficult time conceiving to have children. Given that sperm normally swim around to find an egg and that the first one to make good contact fertilizes it, it is not surprising that the process can be recapitulated in vitro, in a petri dish. Sperm tend to be plentiful and can easily be frozen and stored for later use. While there are some ethical issues with using frozen sperm, these are fairly limited in scope: How much should prospective parents (and offspring) know about a donor? What if the donor dies? Etc. Possible use of surrogacy adds more ethical complexity and makes for popular storylines, but will not be discussed further here.

For IVF, the limiting factor is usually fertilizable eggs. The human ovary contains many immature germ cells, but just one or a few mature each month to be available for fertilization. Hormonal treatments to induce hyperovulation can provide more eggs, making the process more practical and improving the chances of success. But the numbers remain quite small. Technologies for cryopreservation and transplantation of human ovary tissue have been developed to help treat unwanted consequences of menopause. It is not hard to imagine that further developments along these lines might allow for organ culture, and possibly tissue expansion, to make fertilizable eggs both simpler to access and more plentiful. If neither sperm nor eggs are limiting, the issue of choice changes fundamentally as discussed in the end of this section.

## Cloning and Twinning

The current uses of reproductive technologies in humans are quite restricted, but this is not the case for farm animals or animals used as model systems in research. Most people do not consider animals to have individual rights. Manipulation of animal genomes is therefore considered ethically unproblematic, unless it causes excessive suffering or somehow ends up posing a threat to human health. Dolly the sheep captured the imagination and raised public concern because of the implications for use of cloning technologies in humans, not because people lament sheep being genetic copies of one another rather than normal offspring.

Dolly became famous in the 1990s for being a cloned animal, but she was not the first. In 1970, John Gurdon cloned a frog from the nucleus of an adult cell. It only made it to the tadpole stage, but this was a key step in showing that a differentiated adult cell can be reprogrammed to become pluripotent and give rise to a whole animal. The information required, primarily genomic DNA, is present in every cell (with a few exceptions). Many years later, Shinya Yamanaka defined a specific set of proteins, or regulatory factors, that can do the reprogramming, thus generating so-called induced pluripotent stem cells (iPS cells). Gurdon and Yamanaka got a Nobel prize for this work in 2012 (Colman, 2013). After Dolly, many other animals have been cloned, including domestic pets.

The possibility of applying these tools to clone a human is not far-fetched. Reprogramming of adult cells is one possibility, but they may not achieve the required level of pluripotency without side effects. Immature embryonic cells are another option, in which case we may be dealing with twinning rather than cloning. For many mammals, including humans, identical twins occur naturally. The embryo-to-be starts out as a fertilized egg—the first cell with the new, complete genome—which divides to make more cells. In the beginning, the resulting cells are indistinguishable from one another. Normally, they are kept together and cooperate to form one individual. Occasionally, they split apart and, if two separate entities develop, make identical twins. In nature, this happens rarely, and seldom produces more than two identical offspring. In the laboratory, it is possible to deliberately separate cells at the early stage, which can then be used to produce identical twins. Deliberate twinning has been done for cattle, for example (Hashiyada, 2017). In the case of human embryos, cells may be recovered as in the twinning setup or by procedures related to the blastocyst biopsies currently used for genetic testing at IVF clinics. Blastocysts (early embryo cells) and many other cell types withstand freezing quite well, not just sperm. In other words, such cells may be used long after they were collected. If they were to be implanted to produce another child, this child would be an identical twin of the first embryo. In my novel, I call children produced in this way "*youngtwins*", and explore the reasons why parents, or indeed the first-born twin, might choose this route.

It may clarify the ethical debate if we simply think of a clone as an identical twin—a unique individual who has the same genetic makeup as another individual. The potentially more frightening aspect of cloning, namely making many copies rather than just one, will need separate consideration. Embryonic stem cells, which retain most of the properties of a fertilized egg but can be expanded to make a large number of genetically identical individuals, are routinely used to make genetically modified mice. Human embryonic stem cells do exist, but are obviously not used in the same way and may not have exactly the same properties (Ilic and Ogilvie, 2017). Reprogramming of adult cells is theoretically also relevant in this context.

One possible rationale for cloning or twinning is to generate a person as closely resembling an already known person as possible. It could be a uniquely talented person or a much-loved person, possibly deceased. It is easy to imagine complications of "ownership" and decision-making power for frozen embryonic cells. If adult cells are used, this is simplified: Your cells belong to you, not your parents. Ownership is relevant for another possible rationale for cloning—a medical one: A clone could be a body copy of yourself, for spare parts. If we are talking about an actual person, or even a late embryo, this is clearly ethically unacceptable—though the idea has been considered in many speculative fictions. But what if only certain tissues or organs are made? Early embryonic cells can in principle be directed to form anything. There has also been considerable progress in growing tissues from adult-derived stem cells. From simple tissues to functional organs is a significant step. But it is not impossible. Current ideas include the use of human/nonhuman chimeras to host the growth of these cells into fully functional organs for transplantation (De Los Angeles et al., 2018). Chimeras, being partially human, open additional and complex ethical issues.

## Selection and Choice

Choice remains a major—and contentious—issue in human reproduction. Thanks to modern contraception, having sex no longer has to mean the risk of producing a child. Contraception is ethically uncomplicated for many people. This is fortunate on a personal level but also for our planet, challenged as it is by the hugely expanded human population. Abortion is a more complex issue. Our society currently accepts decisions to abort a fetus for one of two reasons: either the woman does not want the pregnancy *and* the fetus has not developed very far—or the fetus has a severe defect, such as a chromosome abnormality or a mutant allele of a gene known to cause disease. With IVF, it is possible to identify embryos with genetic defects beforehand and choose not to implant them. This is accepted and done. We do not currently accept the idea of positive selection of embryos, for example based on sex. In the future, couples wishing to become parents may no longer accept such strict constraints on their ability to choose. Imagine if current bottlenecks of reproductive technology are overcome and a woman could have twenty, fifty or a hundred well-developed eggs available for fertilization. The sperm do their job and there are now many zygotes to choose from to initiate one pregnancy. The future parents might want to avoid alleles linked to high cancer risk or early onset dementia. More controversially, there might be traits that they want to select for. If full genome information is available, it may be tempting to use it to make active choices.

## Reading the Genome

The human genome is large—about 3.2 billion base pairs divided into the 23 chromosomes of a haploid cell. The genome is written in a four-letter code with the bases G, A, T and C. This code is interpreted in many different ways, simultaneously, by the cell. Scientists are very good at reading the parts that encode proteins. This is roughly 1% of the total base pairs, making up a total of 20–25 thousand protein-coding genes. These regions of the genome and the proteins they encode are close to identical in all humans. There are, however, some naturally occurring variants in the population, which may be disease-related or not. In addition, a small number of proteins are hyper-variable, such as those encoded by the major histocompatibility complex (MHC). Proteins are generally very similar in related animals, as well, with a significant fraction being identical in humans and chimps, for example. Other features of the genome—conserved, non-coding genes and certain types of regulatory sequences—are also quite well understood. But there is a lot of DNA in the genome, much of it transcribed into RNA, for which we do not know the function. Some of it may have no function whatsoever: remnants of viruses or transposons or other passive "junk". Much of it, however, is likely to have some regulatory effects, determining how much of a certain gene product is produced in which cells, at which time and in response to which stimuli. In this way, such genomic regions are also part of the code that determines how a complex organ like the brain develops and works. Although regulatory functions are important, they are difficult to decipher directly from the DNA sequence, as they are not read by the cell in as rigid a manner as the protein coding parts are. What *is* clear is that much of the variation between humans and other species, as well as between individual humans, can be found in these less-well-understood areas of the genome. So, while sequencing the first complete human genome—The Human Genome Project—and identifying all the protein-coding genes was an important landmark, sequencing many more individual genomes will continue to be useful for a whole different set of reasons.

One of the practical byproducts of the human genome project is the development of ever faster and cheaper DNA sequencing technology. Until fairly recently, state of the art DNA sequencing meant using the enzyme DNA polymerase coupled with a neat trick of modified building blocks to read longish strings of DNA—typically 500–1000 base pairs. New technologies, called next-generation sequencing (NGS) or massive parallel sequencing, produces a very large number of short sequence-reads at once, which are then assembled and often aligned to a reference sequence by computer (van Dijk et al., 2014). Acquiring the first complete genome sequence of a species is by far the hardest, as it requires fitting together many bits of DNA sequence from scratch and orienting them correctly. Subsequent sequencing of other individuals is much easier because the reference sequence can serve as a scaffold. But, even so, it is remarkable that while sequencing and assembling the first human genome cost about 500 million US dollars and took several years, it now costs about $600 to have your genome sequenced by a commercial sequencing provider and, in principle, takes just one day. A gradual decrease in sequencing costs had been anticipated, but the magnitude and speed of the decrease might not have been. We are now suddenly in the "individual genome era", with all the possibilities and problems that this entails.

So, it is now possible to read the full genome code of every single person on a routine basis—and to store this information. How do we make use of that? One practical application is unambiguous identification, for example for forensics. Given the size of the genome and the presence of variable, even hyper-variable regions, this "fingerprint" can in principle tell each and every one of us apart. Another use of genome-based identification is to assess biological family relationships. This includes straightforward issues like paternity as well as more distant ancestry. The widespread availability of human genome data is having interesting consequences. Some long-sought criminals have been identified through DNA relationships with living relatives. Family trees—in particular the male line—can be reevaluated and unexpected contributions uncovered.

Another use of personal genomes and one that is most relevant for this discussion is identifying gene variants associated with diseases and other traits. Monogenic diseases caused by mutations in a single gene, such as sickle cell anemia or cystic fibrosis, are numerous but relatively rare. Many common diseases, including diabetes, neurodegenerative diseases and various types of cancer, do however have a clear genetic component. Gene variants can be identified that confer high or low risk of that particular disease; they may also correlate with specific treatments being more or less likely to be effective. Disease genes, or more correctly, disease alleles, come with variable degrees of knowledge attached. Sometimes the exact way in which a mutation affects a gene product and a cellular process is known. There may also be direct evidence of causality—for example if the mutation has been studied in an animal model system and correcting the genetic defect was shown to eliminate the disease. At the other end of the spectrum, the link between a DNA sequence and a disease may be based solely on statistics. A sequence variant can be associated with high disease risk without being in any way responsible for the disease—for example by being close to something else in the genome. In that case, fixing the variant would obviously be pointless. In principle, extensive whole genome sequencing allows simultaneous and unbiased correlation of all gene variants with all diseases and traits. It does, however, depend on the appropriate medical diagnostics or trait measurements being performed. As more and more individual genomes are sequenced and analyzed in this manner, even complex associations involving multiple interdependent genes can be uncovered.

Personal genome data give a lot of information about personal risk. As with other rich sources of personal information, this brings forth ethical issues related to privacy, in particular in countries with insurance-based systems for healthcare. From a commercial point of view, statistically demonstrated high risk of disease is just like any other liability: something an insurer will want to avoid taking on if at all possible. Iceland and a Reykjavik-based company, deCODE genetics, has become famous for collecting knowledge about the genomes and genetics of the whole population (Halldorsson et al., 2019). Fortunately, Iceland has a universal healthcare system. Aside from healthcare, there is the question of how much individuals may want to know about disease risks implied by their personal genome, in particular if there is nothing they can do about it.

Finally, returning to the issue of reproduction and choice, genome data for an embryo gives just as much information as for an adult. With sensitive sequencing techniques and prior knowledge of the two contributing genomes (the parents), it should be possible to get this information from one or a few cells. This means that it will be possible to have the complete genome read prior to implantation if using IVF. In other words, even without considering active genome manipulation, there will soon be scope for highly informed and directed choice about which of the potential children to have, should that be desired by parents and allowed by society.

## Manipulating the Genome

As with many technological advances in biological sciences, methods for directed genome manipulation have arisen from scientists studying the basic biology of what cells and organisms normally do and how they do it.

Our genomes are used as operational code in every single cell of our body. This includes very long-lived cells like neurons of the brain as well as short-lived blood cells that patrol the body to catch invaders. Despite the passage of years for long-lived cells, and many rounds of genome duplication in rapidly dividing cells, the genome remains (essentially) the same. Epigenetic changes may affect its readability but the DNA sequence remains. This is not because DNA, as such, is unchangeable. It is easily damaged by UV light and by reactive chemical agents, including endogenous metabolites generated when cells "breathe". Errors also occasionally occur when DNA is copied to prepare for cell division. In normal cells, a dedicated DNA surveillance system detects damage and errors; it then promotes the necessary repairs (Jackson and Bartek, 2009). The genomes of many cancer cells accumulate an excess of mutations over time because this system is defective.

DNA strands can also break, and rejoin. In a controlled manner, this process is responsible for homologous recombination, or exchange, which is used to make the unique haploid genomes of germ cells. It is also required to make the diverse antibodies of the adaptive immune system. Reactive chemical agents can lead to unintentional DNA strand breaks. The cell can repair such breaks using matched DNA from the intact strand or from a homologous chromosome as a template. In addition to the risk of breaking up a gene, a double-stranded break, where both strands of the genomic DNA are broken at once, poses a particular problem because of the way chromosomes are partitioned in dividing cells. For this reason, active mechanisms exist to repair these breaks by rejoining the free ends. This may or may not involve using matched DNA as a template.

In the lab, scientists have found ways to make use of the cellular DNA recombination and repair machinery to deliberately modify the genome. The first generation of this technology involved simply providing cells with a modified exogenous template DNA, which could be used for homologous recombination with the host chromosome and at some low frequency replace the host sequence. The process is quite inefficient, in part due its dependence on a random DNA break to initiate it. Despite this, the existence of mouse embryonic stem cells and optimization of their care have allowed scientists to engineer a large number of genetic changes in mice. This includes inactivating, or knocking-out, individual genes in order to determine their biological role (see for example White et al., 2013). While more challenging, it is also possible to use this method to precisely engineer DNA sequence changes, in other words, to edit the genome. This has often been done to produce mutations in mice that match disease-associated mutations in the human genome and thereby make the most accurate mouse model for that particular disease.

So, directed genome editing is not new; it has been done in mice and other organisms for quite some time. What has changed is the ease with which it can be done. CRISPR/Cas is a bacterial immune system that protects the bacterial genome from invading viruses and plasmids by recognizing their DNA as foreign and cleaving it (Garneau et al., 2010). Genome editing using CRISPR/Cas is powerful and efficient because the CRISPR/Cas enzyme can be targeted to any predefined spot in a genome by the enzyme's RNA component (Jinek et al., 2012). Once there, it will make a break in the DNA, which can then be repaired by the cell. If the goal is just to inactivate the target gene, scientists can take advantage of the naturally occurring non-homologous end joining to create mutations. If a precise editing event is required, exogenous DNA that carries the desired change may be introduced to allow repair via homologous recombination. CRISPR/Cas technology is now being widely used to modify the genomes of cells and model organisms. While it is a major technological advance, there are still some issues with accuracy and unintended consequences that I will discuss in a moment.

Before getting into the safety issues, I would like to point out that there are two different ways in which genome editing technology can be used in human cells. The ethical issues for the two differ substantially. The first and most consequential use would be to edit a fertilized egg or very early embryo. This would, in principle, change every cell in the body of the resulting child and be passed on into future generations. A second use—somatic editing—would be to edit cells from a consenting adult with the intent of introducing them back into to the individual they came from. This could include adult-derived stem cells to correct genetic deficiencies, possibly via in vitro-derived organs, or modified immune cells directed to kill specific cancer cells. If such manipulations cover a significant medical need, one that is currently untreatable or not effectively treated, many people would probably be in favor of it, despite it being a form of human genetic engineering.

The safety issues associated with genome editing must be considered carefully before proceeding with even the most benign-looking application of it. There are two main concerns. The first is accuracy. Basic CRISPR/Cas-mediated DNA editing involves DNA cleavage as well as homologous recombination with exogenous donor DNA. It is known from model systems that either of these processes can lead to unintended changes elsewhere in the genome. More recent method developments, such as base editing and prime editing (Anzalone et al., 2019), allow genome editing without double-stranded breaks or donor DNA and appear to have a lower frequency of unintended changes. In either case, genome sequencing can, and should, be used to check that editing has occurred as planned and that no other changes have been introduced into the genome. Sequencing of the whole genome is essential if the aim is to edit an embryo, but may also be needed for somatic applications.

The second safety concern is unintended consequences of a more subtle kind. This is the tricky bit. Our knowledge of how the genome is used to make a healthy, thinking, feeling person, with all the complexity this implies, is rudimentary. We can make educated guesses about the consequences of altering a gene in humans from studying animal models. But even in well-studied animal models, it is often impossible to predict with certainly what will happen (this will be explored in more detail in the next section). If, however, an allele or variant of a gene is already common in the human population and is not associated with a disease or defect, then it is probably safe to introduce that allele into a specific cell. In practice, this means introducing a normal allele in place of a mutant one, such as CF or BRCA1 mutations, or shifting from one common variant to another. If an allele preexists in the population, we know whether it is basically functional or not. If it is a common allele, we also know how it behaves in combination with many different variants of other genes, which adds confidence. Every other change to the genome is a gamble and it will come down to risk versus reward. For a consenting adult suffering from a devastating disease it may well be worth taking the chance of editing his or her own stem cells and reintroducing them. For reproductive editing, it is much harder to disregard the potential risk, including for subsequent generations. There would have to be a very compelling reason to do it. In the fictional world of *The Unedited*, genome editing is actively pursued because it is the only way to ensure that a particularly vicious virus does not infect people. This scenario is not completely unrealistic, but let us hope it never comes to that.

To close this section, I would like to return to the case of the Chinese twins and the rationale behind that experiment. The twins had one or both of their CCR5 genes inactivated by CRISPR/Cas-9 activity. The stated aim was to make the twins resistant to infection by HIV, which their father carries. CCR5 does encode part of the landing pad for HIV, and people with deletions in their CCR5 genes already exist in the population at about 1% frequency, so the risk of unintended consequences—at this site—could be argued to be limited. Inactivating a gene also requires less accuracy than editing it to make an altered but functional gene product. Thus the approach per se may have technical validity, assuming the twins do not acquire other, unintended mutations as a consequence of editing. By sequencing the twins and their parents, this is in principle knowable (if a bit late). Technical issues aside, other measures could have been used to alleviate the risk of HIV infection. So, not only was the manipulation illegal, it was also non-essential and for that reason not worth the risk of unintended consequences for those two children. Whatever our views on the specifics of this case may be, the fact of it did push human genome editing back into the public discourse—and rightfully so. The possibilities enabled by the new technology may create demand. We need to think carefully about the ethical issues at all levels to allow rational decisions about where society should go with it.

## Predicting and Designing

One of the more emotionally weighted phrases used in discussions about reproductive choice is "designer-babies". Does it reflect anything real? In principle, we could eliminate known disease mutations and switch some other gene variants around. But can we actually design anything, given that we are working with an enormously complex genome that we barely understand? Evolution has created an amazing diversity of plant and animal species displaying a vast array of traits, abilities and peculiarities (Gould, 1980). But evolution is messy; it works with what happens by chance. The genome of any species, including our own, was not designed with every element serving a precise, well-defined function, like in man-made objects. Subtle variations in the genome lead to subtle variations in traits of individuals, but again, not in a predictable, structured way. How do we untangle this?

Leaving aside the issue of design for now, let us start by considering how much we can expect to predict in terms of traits or phenotypes from a given personal genome. By traits I mean both negative traits such as risk of specific diseases and positive traits like strength, beauty, intelligence, musicality etc. The first question is how large the genetic component of a trait of interest actually is, as opposed to the contributions of environment and pure chance. The gold standard for determining this is, or used to be, identical twins reared apart—same genome, different (post-natal) environment. If they share a rare or very specific trait, it probably has a strong genetic component; if they do not, it probably does not. Given how rare identical twins are, and even rarer that they are raised apart, using this comparison as a standard has its limitations. But it is obvious from this approach that certain traits, even if complex, have a very strong genetic component—facial features, for example. Other family studies as well as genome-wide association studies (GWAS) can also give information about the strength of genetic components. My impression is that most traits, including personality and abilities, have *some* genetic component. Just ask a person who is adopted. This does not mean the trait is predetermined, merely that it is influenced by the genetic background. Speaking about disease-risk in terms of genetics is not contentious. Viewing other traits, such as IQ or creativity, in the same manner is not to everyone's liking. But in order to have an informed discussion about future uses of whole genome data and possible genome manipulations, it seems wrong *not* to consider this aspect.

Assuming that there is a genetic component to a trait, how easy is it to identify the contributing genes? This depends on the magnitude and the complexity of the genetic contribution—the larger it is and the fewer genes involved, the higher the predictive power. The two are not necessarily linked: Eye color is genetically simple, facial features are not, but both have a strong genetic component. It also depends on the number of individuals who have had their DNA analyzed, ideally their genomes sequenced, *and* this trait reliably measured. Genome sequencing is expanding and the data are accumulating rapidly. Trait measurements are more variable. Physical characteristics are the most straightforward, along with well-defined disease states. But even these tend to vary over time and require an objective observer or an objective test. Certain tests, like IQ measurements that give a number, seem straightforward but are not. People can improve their scores by training; that does not mean they become smarter. Finally, measurements and genomes need to be reliably paired. For diseases, doctors, clinics and lab-tests are responsible for objective diagnoses and patients generally cooperate because it is in their best interest. But wider use of this information requires good recordkeeping and external access to the data, which has ethical complications as discussed previously. Keeping genetic tests and diagnoses connected but dissociated from patients' names or ID numbers can in principle help maintain anonymity. But as we shift toward using complete genome sequences as DNA data, this may break down. The personal genome is, after all, the ultimate identifier. In conclusion, if a trait has a high genetic component *and* it is properly recorded, genome sequencing may allow reliable prediction of that trait in the future, even if it cannot do so now. It will, however, remain a matter of statistics and rarely a certainty.

Can we predict the consequences of deliberately altering the genome? That is, can we predict the impact of introducing a specific change in the context of a specific, personal genome? Can we re-design a person? That depends on what is meant by design. Correcting mutations by reintroducing wild type alleles or substituting variable alleles may soon be possible with existing technology, has limited risks and fairly predictable outcomes. But this stays within the range of naturally occurring variation between humans and hardly qualifies as design. What about *real* changes? By this I mean introducing changes to the genome that are *not* already present in the population. We can call these genetic novelties. Such changes might be tempting if they seem to be extremely favorable. Imagine inactivating a gene that is directly involved in the development of Alzheimer's or other dementia; imagine a simple mutation that is likely to increase active lifespan. Making more drastic changes to our biology, what we might call *de novo* design, seems unrealistic on the pure biology side. We simply do not understand genome logic and its translation into real-life biology well enough to design reliably that way. One type of "design" merits a separate mention: copying. In the case of a delayed identical twin or a clone, we have a pretty good idea what the outcome would be. Twinning or cloning humans is likely to become technically feasible, in one form or the other, in the near future. The question is whether there will ever be public interest in making it legal. Another route that limits possible unforeseen consequences is what I previously termed somatic manipulations. Any change is restricted to the particular cells being manipulated and the ethical concerns are minimized as only the person agreeing to the treatment is affected. Such genome editing approaches seem likely to move forward and may eventually become quite sophisticated.

Any genetic novelty that one considers introducing into the genome of a human embryo should be checked in a model system first. This seems reasonable. One of the reasons we study genetics of model systems such as mice, fruit flies and roundworms is to understand the contribution of specific genes and gene products to the function of the organism as a whole. No cell line, tissue or even whole organ grown in culture can address this adequately. The genome did, after all, evolve as a blueprint for the whole animal, not for isolated cells. Genetics continuously throws up surprises, however, even in the best-studied model systems. This is true for forward genetics, where scientists look for a gene responsible for a trait/phenotype of interest, as well as for reverse genetics, which looks at the consequences of manipulating a pre-defined gene. An example of the first kind is the discovery of a whole new class of non-coding genes, micro-RNAs, from changes in the timing of differentiation events in the roundworm (Lee et al., 1993). An example of the second kind is inactivating the mouse gene encoding uPA (urokinase-type plasminogen activator). Interest in uPA comes largely from its association with cancer progression and tissue remodeling. The mutant mouse, however, is timid: it shows reduced exploratory activity (Rantala et al., 2015). In fact, many mouse mutants have unexpected phenotypes (White et al., 2013). Also, the same mutation introduced into different inbred mouse strains can have surprisingly different effects (Threadgill et al., 1995). In humans, this corresponds to the same variant of a gene having different effects depending on the genetic background, that is, which regulatory landscape and which variants of other genes are present in a person's genome. Such interdependencies may turn out to be the rule rather than the exception for naturally occurring mutations and the traits that we care about.

In conclusion, the effects of genetic perturbations are often not what we would expect them to be. Add to this that any change made to the genome of a zygote will be present in each and every cell of the body, including the brain, and we face what may be an insurmountable challenge. No matter what model system is used for testing, it will never be possible to ensure that a genetic novelty will not trigger unintentional neurological or psychological effects—psychosis, depression, inability to learn syntax of a language or some other quirk of the brain. We cannot ask mice or flies—or even monkeys—how they feel. Nor could we trust their answers to reflect how we would feel.

## Concluding Remarks: Choice Versus Chance

Technological advances give us possibilities. Fast, inexpensive whole genome sequencing is one such advance. Simply having this information, in particular for that little ball of cells that could produce a whole new person, presents any number of challenges from an ethical perspective. It is all about choice. The possibility of manipulating—or editing—the genome brings even more possibilities and more choices. We, as individuals or as a society, will be responsible for the consequences whether we act on the choices or not.

A fictional example illustrates how challenging this can be: If, based on a large-scale IQ test, it is found that individuals with the allele combination a4a4, b2b2, c1c4 have a 40% of having an IQ > 145 (exceptionally gifted) whereas for a4a4, b2b2, c1c1 it is 0%, would you be tempted to implant a c1c4 embryo over a c1c1 embryo if given the choice? Or would you decline to know and effectively toss a coin? What if the choice is about 0% versus 40% chance of having an IQ < 35, that is, severely impaired and likely to need lifetime care? Would you then choose the c1c4 embryo? Would society always pay for care if you didn't? In either case, would you choose to edit a c1 allele to c4 (assuming it is technically safe) for an embryo? What if we substitute IQ for creativity or freedom from depression or low cancer risk? What if you choose not to know or not to edit and the affected child realizes this later on?

In the novel *The Unedited,* I explore questions like these via the fictional lives of five young characters. This is not just future and fiction, however: fertility clinics are already offering the choice to select against certain genetic variants. Currently these choices are limited, but then sequencing the whole genome has also only recently become easy and inexpensive.

Choice versus chance—that is the question. In most areas of life we think choice is best. In reproduction, having children, it gets complicated. A new child is an unknown, a person created by two people plus a hefty dose of chance; this seems natural and comfortable to us. Some parents-to-be even prefer *not* to know the sex of their child before birth. So, it is possible that there will never be any public pressure to make complete genome information available and allow informed choices—beyond what is considered critical for a child's wellbeing. Nazi eugenics casts a long shadow in our culture. But attitudes may change as genome-based predictions become more sophisticated and as new diseases or new viruses arise. We are already close to this dilemma: Is it ethical to bring a child into this world if she has 70% chance of breast cancer or early-onset Alzheimer's or severe depressions? Also, people differ. Some parents are likely to want as much information and control as possible. IVF clinics offer some DNA screening of pre-implantation embryos; sending the DNA to a whole genome sequencing facility is a small step. These questions and issues are unlikely to ever be easy, for the individual or for society. But they will, hopefully, be at the center of a sensible debate moving forward.

# Bibliography

Anzalone, A. V. et al: Search-and-replace genome editing without double-strand breaks or donor DNA. Nature. doi:https://doi.org/10.1038/s41586-019-1711-4 (2019)

Colman, A.: Profile of John Gurdon and Shinya Yamanaka, 2012 Nobel Laureates in Medicine or Physiology. Proc Natl Acad Sci U S A. **110**(15), 5740–5741 (2013)

Cyranoski, D. and Ledford, H.: Genome-edited baby claim provokes international outcry. Nature. **563**, 607–608 (2018)

De Los Angeles, A., Pho, N. and Redmond, D. E. Jr.: Generating Human Organs via Interspecies Chimera Formation: Advances and Barriers. Yale J Biol Med. **91** (3), 333–342 (2018)

Garneau, J.E., Dupuis, M.È., Villion, M., Romero, D.A., Barrangou, R., Boyaval, P., Fremaux, C., Horvath P., Magadán, A.H. and Moineau, S.: The CRISPR/Cas bacterial immune system cleaves bacteriophage and plasmid DNA. Nature. **468** (7320), 67–71 (2010)

Gould, S. J.: The Panda's Thumb. Publ.: New York: W. W. Norton. (1980)

Halldorsson, B.V. et al.: Characterizing mutagenic effects of recombination through a sequence-level genetic map. Science. **363**, (6425), eaau1043 (2019)

Hashiyada, Y.: The contribution of efficient production of monozygotic twins to beef cattle breeding. J Reprod Dev. **63** (6), 527–538 (2017)

Ilic, D. and Ogilvie, C.: Human Embryonic Stem Cells-What Have We Done? What Are We Doing? Where Are We Going? Stem Cells. **35** (1), 17–25 (2017)

Jackson, S.P. and Bartek, J.: The DNA-damage response in human biology and disease. Nature. **461** (7267), 1071–8. (2009)

Jinek, M., Chylinski, K., Fonfara, I., Hauer, M., Doudna, J.A, and Charpentier, E.: A programmable dual RNA-guided DNA endonuclease in adaptive bacterial immunity. Science. **337**(6096), 816–821 (2012)

Lee, R. C., Feinbaum, R. L. and Ambros, V.: The *C. Elegans* heterochronic gene lin-4 encodes small RNAs with antisense complementarity to lin-14. Cell. **75** (5), 843–854 (1993).

Rantala, J., Kemppainen, S., Ndode-Ekane, X.E., Lahtinen, L., Bolkvadze, T., Gurevicius, K., Tanila, H. and Pitkänen, A. Urokinase-type plasminogen activator deficiency has little effect on seizure susceptibility and acquired epilepsy phenotype but reduces spontaneous exploration in mice. Epilepsy & Behavior. **42**, 117–28. (2015)

Threadgill, D. W. et al.: Targeted disruption of mouse EGF receptor: effect of genetic background on mutant phenotype. Science. **269** (5221), 230–234 (1995)

van Dijk, E.L., Auger, H., Jaszczyszyn, Y. and Thermes, C.: Ten years of next-generation sequencing technology. Trends in Genetics. **30** (9), 418–26 (2014)

White, J.K. et al.: Genome-wide Generation and Systematic Phenotyping of Knockout Mice Reveals New Roles for Many Genes. Cell. **154** (2), 452–464 (2013)

Printed in the United States
By Bookmasters